FOREWARD i
ABOUT THE AUTHOR i
INTRODUCTION 1
DESIGN FACTORS 1
ROOF DECKS 5
DRAINAGE 11
ROOF INSULATION 14
MEMBRANE MATERIALS 25
FLASHING 32
IN THE FIELD 51
ABBREVIATIONS 59
GLOSSARY OF TERMS 61
BIBLIOGRAPHY 69
RESOURCES 71
QUIZ 73

Low-Slope Roofing I

National Council of Architectural Registration Boards
1735 New York Avenue, N.W., Suite 700
Washington, DC 20006
(202) 783-6500
http://www.ncarb.org

ISBN 0-941575-31-4
Printed in the United States of America

This monograph was published in May 2000.

FOREWORD

This is the seventh monograph in NCARB's Professional Development Program (PDP). This program is intended to help architects meet mandatory continuing education and professional development requirements of state registration boards. It is also accepted by The American Institute of Architects (AIA) to meet its continuing education requirements for maintenance of membership in the Institute.

NCARB's PDP Committee selects topics, reviews content and prepares tests for the monographs. During the past year, two new topics have been under development: sustainable design and professional conduct for architects.

This monograph is the first of two on low-slope roofing. Written for architects, students and others, it provides basic information to meet the design performance required of built-up roofing (BUR) assemblies. The second, to be released in the future, will address single-ply, modified bitumen, and special roofing materials.

NCARB welcomes your comments and questions about this monograph and about the Professional Development Program in general. Please let us know what other topics you would like to see addressed. Thank you for your participation.

ABOUT THE AUTHORS

Justin Henshell, AIA, NCARB, is a partner at Henshell & Buccellato, Consulting Architects, in Red Bank, New Jersey. A fellow of the American Society for Testing and Materials (ASTM), he serves on Committees D-08 Roofing & Waterproofing, C-15 Masonry Units and E-06 Performance of Building Constructions. He is also chairman of Subcommittee D-08.20 Roofing Membrane Systems. In addition, Henshell is a member of Working Commission W86 Building Pathology of the International Council for Building Research Studies and Documentation (CIB).

Henshell is a current faculty member and past regent of the Roofing Industry Educational Institute (RIEI). He lectured on BUR in continuing education workshops sponsored by the University of Wisconsin's Department of Engineering. He has also lectured on roofing, flashing, masonry and other construction materials at other universities, trade organizations and professional settings. Henshell has written technical papers for a variety of academic and professional organizations, including "Specifying Membrane Roofing" and "Built-up Roofing Samples" for, respectively, the November 1987 and February 1995 issues of the *Construction Specifier*. He is also co-author of "Precast Concrete Pavers for Parking Decks," which appeared in ASTM's May 1990 publication entitled *STP 1084*.

A registered architect, Henshell is licensed in six states and holds a certificate from NCARB. He is a member of the AIA, the New Jersey Society of Architects, the American Concrete Institute (ACI) and the Construction Specifications Institute (CSI). He formerly served as president of the CSI Metropolitan New York chapter.

Donald E. Brotherson, AIA, is a consulting architect in private practice in Aventura, Florida. He is a fellow of ASTM and past chairman of Committee D-08 Roofing & Waterproofing. He is a former member of ACI's Committee C515 Waterproofing of Concrete Structures and serves on CIB's Working Commission W86 Building Pathology.

Brotherson is a current faculty member of RIEI. In the past, he has also served as a regent and chaired its original course curriculum committee. He has lectured on BUR in continuing education workshops sponsored by the University of Wisconsin's Department of Engineering. He also has lectured at other universities, trade organizations and professional settings on roofing, maintenance management, energy conservation, and mechanical systems, among other construction topics. In addition, he has written technical papers on building pathology, specifically in regards to roofing and waterproofing.

A registered architect in Illinois and Florida, Brotherson is a member of the AIA. He is also Research Professor of Architecture Emeritus and former director of the Building Research Council at the University of Illinois.

Paul Buccellato, AIA, NCARB, is a partner at Henshell & Buccellato, Consulting Architects, in Red Bank, New Jersey. He is a member of ASTM's Committee D-08 Roofing & Waterproofing, serves as vice chairman for Subcommittee D-08.22 Waterproofing & Dampproofing and secretary for Subcommittee D-08.20 Roofing Membrane Systems, and chairs task groups on low-temperature unrolling of felt and EPDM flashing details. He is also a member of C-15 Masonry Units.

Buccellato currently serves as a regent of RIEI. He has lectured in the Architectural Technology Department of Brookdale College in Lincroft, New Jersey, and has authored technical articles that have appeared in various trade publications. He co-authored "Precast Concrete Pavers for Parking Decks," noted above. He has written and co-authored several technical papers, including "Detailing Flashings in Single-Ply Roofing Systems," which appeared in the October 1989 issue of the *Construction Specifier*. In addition, Buccellato is a contributing author to *The Roofing Specifier*.

A registered architect in New Jersey, New York, Pennsylvania and Virginia, Buccellato holds a certificate from NCARB. He is a member of the AIA, the New Jersey Society of Architects, and the CSI. He is a past member of the American Arbitration Association. He is also vice chairman of the planning board in Matawan, New Jersey.

INTRODUCTION

The roof assembly is one of the most important components of a building envelope. It must be designed to handle a number of tasks. A roof must keep a building dry, of course, but it also must be able to support dead loads such as mechanical equipment; carry live loads such as pedestrian traffic, rain, snow and wind; and resist fire and thermal transmission. Some roofs provide bearing surfaces for recreational activities and others serve as platforms for signage and even helicopter pads.

Because many of these functions are critical to the satisfactory performance of a building, the roof should be considered from the inception of building design. A roof that is merely an afterthought is an invitation to disaster and litigation. In fact, roofing problems generate more litigation than any other building element. In "Roofing: Synthesizing Design and Craftsmanship," which appeared in the February 1987 issue of *Architectural Record, Engineering News-Record* Senior Editor Peter Green noted that "Roofing may be one of the least rewarding aspects of an architect's design program, but it deserves undivided attention or else it can become the most conspicuous of the program's shortcomings. (Unfortunately, buildings are often remembered in the industry, not for their design awards, but for their roofing litigation)."

This monograph addresses hot applied built-up roofing (BUR), a type of low-slope roofing system specified on about 35 percent of new buildings in the U.S. today. BUR, which has been in use since the mid-1800s, is a continuous roof covering consisting of plies of saturated or coated felts alternating with layers of bitumen, surfaced with aggregate, mineral granules or asphalt. Other low-slope roofing systems, such as single-ply or modified bitumen membranes, are not covered here. Nor is reroofing included, although many of the principles discussed in the following pages apply to the replacement of existing roofs. The information presented in this monograph is relevant to the new, so-called hybrid systems, which consist of multiple plies of felts and bitumen covered by a modified bitumen cap sheet. (For more information on hybrid systems, refer to NCARB's upcoming monograph, *Low-slope Roofing II.*)

The monograph discusses the entire roof assembly: roof deck, vapor retarder, insulation and membrane. Specific topics include code requirements, design considerations, material options, and installation techniques. Roofing applications for areas subjected to high winds are not discussed in detail.

While reading and successfully completing the questions at the end of this monograph may not transform an architect into a roofing consultant, it will surely help a practitioner become more familiar with roofing terminology and concepts and more skilled in designing BUR systems.

DESIGN FACTORS

WHY SPECIFY BUR?

BUR has been successfully applied for more than 100 years. The crude systems of the 1840s, consisting of pine-tar treated paper, have evolved into sophisticated assemblies that feature, for example, very strong glass-fiber felts. Although problems can—and do—occur with BUR applications, experience indicates that these assemblies perform well when care has been given to their design, installation and maintenance. Both asphalt and coal-tar pitch—the two types of adhesives used in

BUR—demonstrate excellent resistance to water; they work well with other building materials; and, being basically thermoplastic, they can stretch and move to accommodate movements in the roof assembly.

BUR, used in about one-third of all roofing applications, has held up well against its competition. Of the many single-ply systems that were brought into the market in the early 1980s, only a few remain and even they have had to be improved several times in order to perform adequately.

OCCUPANCY

When designing a roof assembly, an architect must consider the building's occupancy and use. Corrugated metal panels may suffice to cover a warehouse for farm equipment, where an occasional leak could be tolerated. However, structures housing sensitive equipment or functions that cannot afford water damage may require very sophisticated roofing systems with several redundant layers to maintain their watertight integrity.

BUILDING CODE AND INSURANCE REQUIREMENTS

Roofing systems, like all building components, must be designed to meet local building codes and satisfy insurance requirements. It is critical that architects check codes and insurance company requirements before selecting a roofing system.

Code Requirements

Most local codes are based on one of three model building codes. Efforts are being made to develop one model code but it has not been adopted as yet. The model building codes generally follow or reference tests and standards established by ASTM or Underwriters Laboratories, Inc. (UL). In regard to roofs, building codes are primarily concerned with flame spread on the building exterior and structural failure.

UL and other test references provide tests for three different fire conditions:

- External fire exposure: Assemblies are rated as Class A, B or C, depending on their ability to resist fire on top of the roof. Tests for this condition are ASTM E108 (Test Methods for Fire Tests of Roof Coverings) or UL 790 (Tests for Fire Resistance of Roof Covering Materials).
- Internal fire spread: a measure of the time it takes for fire or flame to spread under the roof deck. The appropriate test is UL 1256 (Test Method of Roof Deck Construction).
- Internal fire resistance: a measure of the endurance of the roof assembly in terms of the time it takes for the assembly to collapse. Tests for this condition are ASTM E119 (Test Methods for Fire Tests of Building Construction and Materials) or UL 263 (Test Method for the Hourly Internal Fire Resistance of Roof-Ceiling Assemblies).

The fire tests are performed on a complete roofing assembly, not on separate components, so the entire system must be specified to earn the classification. Note that the classifications are concerned with fire resistance only, not with how well the roof assembly performs in general. Also, building materials delivered to the construction site should carry the appropriate UL labels to assure that they have been fabricated from the same materials used in the various tests.

Insurance Requirements

The architect must also be aware of the requirements of the building owner's insurance provider. Insurance companies are concerned with the spread of internal flame and wind damage that will affect the contents of their insured buildings. Assembly requirements can be found in *Factory Mutual System: A Guide to Equipment, Materials & Services Approved by Factory Mutual Research for Property Conservation* and *Loss Prevention Data Sheets*, both published by Factory Mutual Research Corporation (FM), an organization of mutual property insurance companies that acts as an advisor to insurance underwriters and building owners. Its laboratory establishes standards and verifies compliance.

In terms of fire, FM's primary requirement is limiting its spread on the underside of the roof assembly. The organization divides roof assemblies into two classifications: unsprinklered and sprinklered. Class 1 steel-deck assemblies, noncombustible decks, and treated wood decks do not require sprinklers. Combustible and Class 2 steel decks require sprinklers. Whether a steel deck is Class 1 or 2 depends upon the heat release—the amount of heat generated when subjected to fire—of the assembly's components.

High winds blowing over a building can create strong enough forces to lift a roof assembly or components of the assembly. This is referred to as *blow-off*. An architect will need to look at design criteria stipulated by UL, FM, or the American Society of Civil Engineers (ASCE), depending on the specific requirements of the project and local building codes. A structural consultant can help an architect decide what wind resistance is needed for a particular roofing assembly.

UL classifies wind-uplift resistance as Class 30, Class 60 and Class 90, depending on the ability of the roof assembly to resist 45, 75, or 105 pounds per square foot (psf) negative pressures in the UL wind-uplift test.

FM Loss Prevention Data Sheet 1-128 contains all of the needed data to determine wind loads and roof deck securement. *Data Sheet 1-128* rates assemblies according to classifications known as 1-60, 1-90, 1-120, 1-150 and 1-180. The number following the hyphen denotes resistance in pounds per square foot for a one-minute period during the FM wind-uplift test.

ASCE and some local building codes may differ from FM in how they calculate wind-uplift loads but the results are comparable.

DURABILITY

Most roofing systems are expected to provide a service life of 10 to 20 years. BUR has an excellent history of meeting and exceeding that expectation when properly designed, applied, inspected and maintained. Unfortunately, poor roof design, installation and in-service abuse can severely shorten the expected life; roofs are frequently replaced in as little as 4 or 5 years.

A roof's expected life should be matched to the building's current use or expected service life. An architect may not be able to justify the expense of a very durable roof for factories and other buildings that typically undergo several changes of use during their service life since changes in occupancy and interior conditions may warrant changes in the roof assembly as well. On occasion, a building may only be temporary, with demolition expected in a short time, so a roof assembly with a 5- or 10-year life expectancy may be more appropriate than one with a 20-year life.

Durability must also be balanced out with other qualities: sometimes, for example, reflectiveness or cleanability may have a higher priority.

APPEARANCE

Most BUR systems are out of sight and therefore do not present an aesthetic consideration. But those on a lower building can sometimes be seen from above. In that case, the selection of surface treatments becomes significant and more expensive coatings, reflective aggregates and surfaced sheets may be desired instead of standard gravel or crushed stone surfacings. Decorative surfaces can be created with varied colored aggregates but it is usually at the cost of durability or maintenance.

ENVIRONMENTAL FACTORS

Roof assemblies must withstand the most severe conditions imposed on a building. Temperatures on a given roof surface can vary from lows of -50 F [-45.5 C] to highs of 180 F [82.2 C], depending on the location of the building, the color of the roof and the mass of the roof structure. In systems with high insulation values, these temperature extremes are even more pronounced.

Temperature changes within a roof assembly cause differential movement of the components because different materials expand and contract at different rates. Differential movement can rupture weaker components, such as roofing felts.

Wet climates also can contribute to roofing problems. Moisture that has been built in or trapped within a roofing assembly can cause blisters, ridging and buckling of the roofing felts. And moisture within insulation reduces thermal efficiency. Environmental factors are regional. Membranes in Denver, for example, require a high degree of resistance to ultraviolet (UV) radiation whereas those in Miami must resist hurricane forces.

WARRANTIES

A manufacturer's warranty is backed by the company and not by an insurance company or other third party. As such, they are only as good as the company. If the company closes or leaves the country—as many have—the warranty can be useless. Warranties should be evaluated in terms of what they will return to the owner for the cost of the warranty. The warranty establishes a direct relationship between the building owner and the manufacturer of the roofing materials. It defines the manufacturer's responsibilities and obligations to the owner, irrespective of the specifier or applicator. In most cases, it limits the manufacturer's obligations rather than extends them.

In its annual guide to commercial roofing products, the National Roofing Contractors Association (NRCA) provides a listing of most of the warranties available. The list includes a summary of the items covered by the warranty, monetary limits, exclusions and some twenty other aspects of the warranty. The architect should be well aware of the limitations of the warranty before it is specified.

Applicator warranties are generally provided for a 2-year period after the roof membrane application is accepted. It is possible to negotiate or specify longer warranty periods. However, the warranty is only as good as the company that backs it and only as long as the company is in business.

LIFE-CYCLE COSTING

Life-cycle costing and value engineering are sciences unto themselves, well beyond the scope of this monograph. Architects who provide this type of service should consult with experts in the field before presenting such information to their clients. The cost of money, expected service life, maintenance costs, replacement costs, and present value are only a few of the many variables to be considered.

LOGISTICS

All building components have their advantages and disadvantages. BUR systems are excellent roofing systems but they may not be appropriate on a small roof 40 stories high in a congested urban area. BUR components are heavy and bulky, very hot when applied and, in most cases, fairly dangerous materials with which to work. They also give off unpleasant fumes that may not be acceptable in all areas. Typically, a generous amount of space is needed on the ground to park vehicles and equipment associated with BUR applications (figure 1). All of these factors must be evaluated when selecting a roofing system.

FIGURE 1
BUR LOGISTICS
Ample space is required at ground level.

ROOF DECKS

GENERAL

The roof deck is the platform that supports the insulation and membrane. The following are some of the attributes required of a roof deck:

- structurally sound
- stiff
- well secured to the roof's structural framing system
- fire resistant to the extent required by codes
- dimensionally stable in the presence of moisture
- resistant to wind uplift
- capable of receiving and retaining fasteners.

A roof deck must be capable of safely supporting the loads mandated by local building codes or the increased loads specified by the designer. In addition to its own weight, a deck structure must be able to support the weight of the following:

- the roof system, including insulation, membrane and surfacing
- dead loads of mechanical equipment, penthouses and bulkheads
- environmental loads, such as rain, snow, wind and seismic forces
- transient loads from rooftop traffic.

A roof deck and its structural supports will deflect under these loads. Concrete and wood decks can also deflect considerably due to creep. When a deck system deflects, a depression that can trap water forms between supports. The additional weight of this water will cause additional deflection. Reasonable limits on deflection are necessary to maintain positive drainage, thereby preventing water from ponding.

Dimensional stability of the deck depends on its coefficient of thermal expansion and contraction and, in the case of organic fibrous material, its degree of swelling due to moisture absorption. Excessive deck movement can wrinkle or split a BUR membrane. To prevent delamination of the assembly by such horizontal movement or by wind uplift, all roof components must be positively secured.

TYPES OF DECKS

In this section we will discuss the most common types of roof decks used throughout the United States (figure 2). Decks are categorized by their resistance to fire (combustible or noncombustible) and their ability to receive and retain fasteners (nailable or non-nailable). Roof systems installed over a nailable deck are secured with mechanical fasteners. Non-nailable decks require the system to be installed in hot bitumen or cold adhesive.

FIGURE 2

Types of Roof Decks

	Noncombustible	Combustible
Non-Nailable	Cast-in-place concrete Precast concrete Prestressed concrete	
Nailable	Steel Lightweight insulating concrete Precast or poured-in-place gypsum concrete Cementitious wood fiber	Wood Plywood

Noncombustible, Non-Nailable Decks

These systems include cast-in-place concrete; precast concrete planks and channels; and prestressed tees, double tees and cored planks. They are all constructed of standard weight structural concrete (150 pounds per cubic foot, or pcf [2403.0 kg/m^3]) or lightweight structural concrete (110 to 115 pcf [1762.2 to 1842.3 kg/m^3]). These decks are fire resistant and can offer an attractive, finished ceiling.

Mechanical fasteners are difficult and costly to use in concrete, so insulation and roof membrane are usually attached to these decks with hot bitumen. Concrete decks should be primed before installation of a vapor retarder, insulation or membrane. The primer absorbs the surface dust on the concrete.

Deck surfaces must be free of sharp ridges, frost, moisture, debris or coatings that would inhibit bonding of the roof system. Sharp ridges can cut through a membrane applied directly to the deck. Frost or moisture will cause foaming of the hot bitumen, thereby preventing an adequate bond between the deck and roof system. Some concrete curing agents will also inhibit the adhesive bond, so architects must be careful as to which products are specified.

Cast-in-Place Concrete

Cast-in-place concrete can be sloped to drain during placement and provides an excellent substrate for a roofing system. The deck must be cured and dry prior to roofing. Surfaces must be relatively smooth and flat to permit bonding of the roofing components. The deck surface should be no rougher than a wood float finish.

Precast Planks and Channels

These members are relatively short in length (less than 10 feet [3.04 m]), narrow in width (16 to 24 inches [406.40 to 609.60 mm]), and shallow in depth (2 to 4 inches [50.80 to 101.60 mm]). They are usually installed on steel joists

FIGURE 3
Precast Concrete Plank

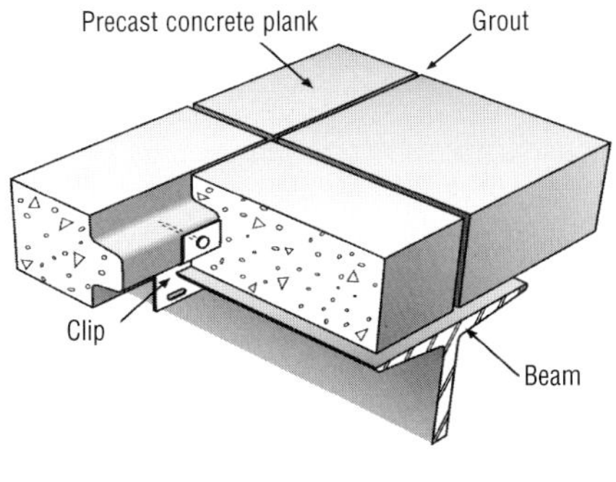

and beams with clips (figure 3). Planks are subject to movement caused by vibration and by changes in temperature and moisture. Because the clips do not offer sufficient movement resistance, the joints between planks are filled with grout. A base sheet is usually adhered with strip mopping or the joints are covered with a 6-inch-wide [152.40 mm] fiberglass strip, then the insulation is adhered in hot asphalt.

Precast and Prestressed Concrete

These decks include single tees, double tees (figure 4), and cored prestressed planks (figure 5). They vary in width from about 2 to 8 feet [.61 to 2.44 m]. With spans of up to approximately 60 feet [18.29m], they can provide column-free spaces and accommodate cantilevered designs. They do not, however, provide a satisfactory surface to receive roofing components. Prestressing induces camber in the concrete member to counter some of the subsequent deflection that develops after it is installed and loaded. Unfortunately, the camber is rarely uniform, resulting in offsets between adjacent members. A 2-inch-thick [50.80 mm] cement topping should be applied to create a smooth flat surface and cover the lifting hooks and weld plates. The topping thickness can be varied to provide a positive slope to drain.

FIGURE 4
Precast Concrete Double Tee

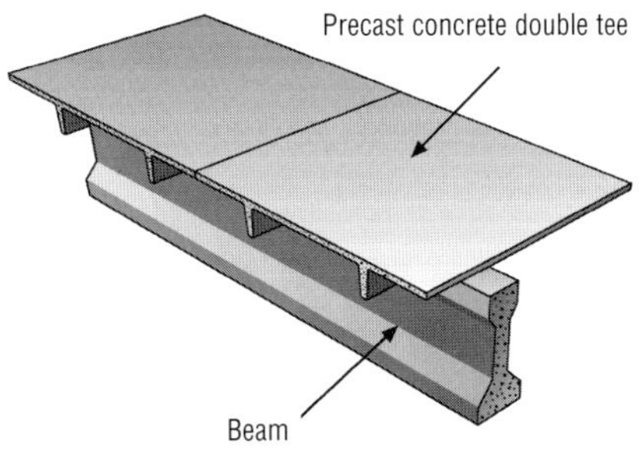

FIGURE 5
Cored Prestressed Plank

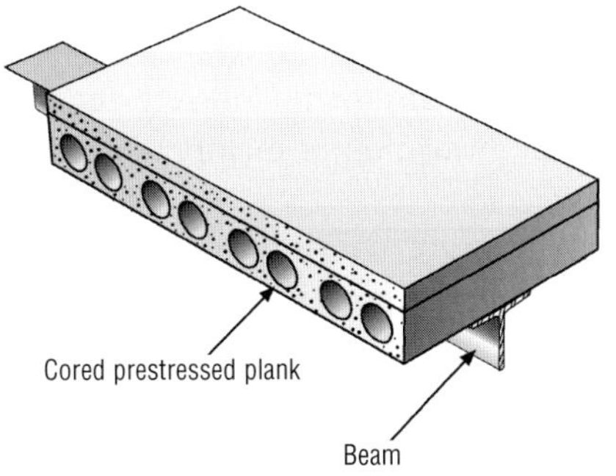

Noncombustible, Nailable Decks

These include steel, lightweight insulating concrete, gypsum concrete, and cementitious wood-fiber decks. They may also include precast concrete plank when the units are cast with aggregates that can receive fasteners.

These decks offer several advantages, including fire resistance, some insulating qualities, and lighter weight than non-nailable systems. Resistance to wind uplift is achieved by mechanical fasteners.

Except for steel, these decks incorporate water, either as a mix or as part of a grouting system, at the time that they are installed. The deck may not be dry at the time of roofing. Moisture will inhibit bonding with hot materials and subsequent shrinkage and cracking may rupture the membrane. Mechanically fastening a base sheet to the deck lessens the chance of developing problems associated with moisture, cracking or shrinkage.

The surfaces of these decks must be reasonably smooth, dry and free of projections and ridges. Surface irregularities must be leveled and hollows filled. Priming is not required, since the base sheet will be mechanically fastened to the deck.

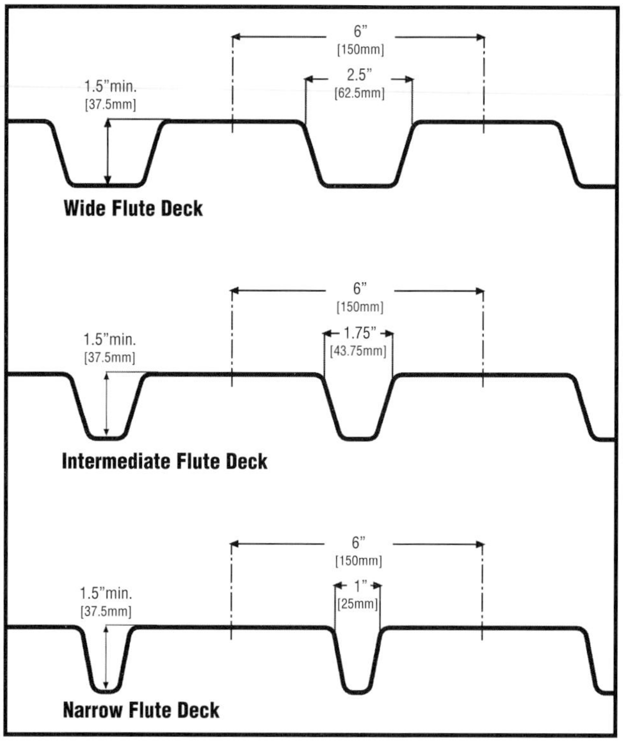

FIGURE 6
Types of Steel Decks

Steel

Steel is the most widely used roof deck because it is lightweight, fire-resistant, easy and fast to erect and can double as a diaphragm. Design information can be obtained from publications issued by the Steel Deck Institute (SDI) and FM.

Decks are roll-formed from painted or galvanized steel sheets. Ribs formed in the sheets provide strength and rigidity. The top surface is called the flange and the ribs are called flutes. Steel decks are usually 1-1/2 inch [38.10 mm] deep.

Steel decks are identified according to the width of their flutes (figure 6):

- narrow flute: 1 inch [25.40 mm] wide
- intermediate flute: 1-3/4 inch [44.45 mm] wide
- wide flute: 2-1/2 inch [63.50 mm] wide.

The most common thicknesses are 22 [0.71 mm], 20 [0.86 mm] and 18 [1.14 mm] gauge measured on the uncoated steel. Twenty gauge is used for long spans and where the deck serves as a diaphragm. Select the gauge to limit deflection to 1/240 of the span.

Decking is installed over steel framing and secured 12 inches [304.80 mm] on center (maximum) by welding or fastening with self-drilling screws. Welds should be ½-inch [12.7 mm] diameter. End laps should be 2 inches [50.80 mm] and occur over supports. Side laps, occurring at least 36 inches [914.4 mm] on center, should be secured with self-tapping screws. Deck openings should be framed to support the deck and minimize deflection.

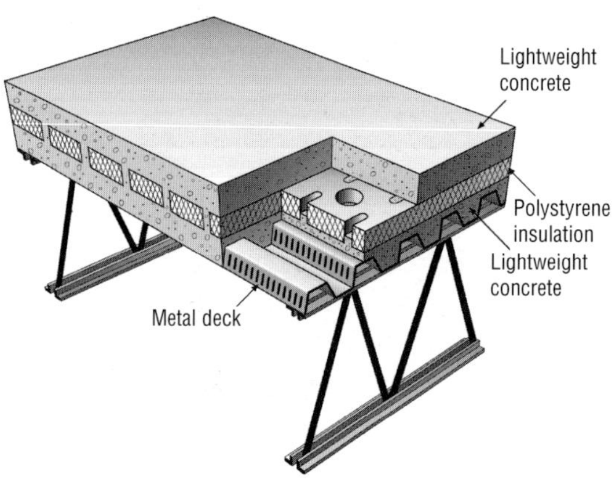

FIGURE 7
Lightweight Insulating Concrete Deck

Lightweight Insulating Concrete

Because of their lighter weight and greater thermal resistance, lightweight insulating concrete decks (figure 7) have supplanted gypsum decks in all but a few southern states. They are made by combining portland cement with either lightweight aggregates, such as perlite or vermiculite, or with a liquid foaming agent.

The usual mix is 1:6, portland cement to aggregate, but can vary from 1:8 to 1:4. The latter is used with reinforcing where seismic resistance is required. The average density is 22 pcf [352.44 kg/m^3].

Insulating concrete systems should be applied over venting galvanized steel decks to permit downward drying. The material can also be installed over cast-in-place or precast structural concrete decks. To reduce moisture content in these cases, it is mixed in the ratio of 1:4 and installed in thickness of no more than 2 inches [50.80 mm] over slotted expanded polystyrene insulation.

Because these decks can deteriorate in environments of high relative humidity, they should not be used in pools, kitchens, or other areas where the relative humidity (RH) will exceed 30 percent.

FIGURE 8
Gypsum Concrete Deck

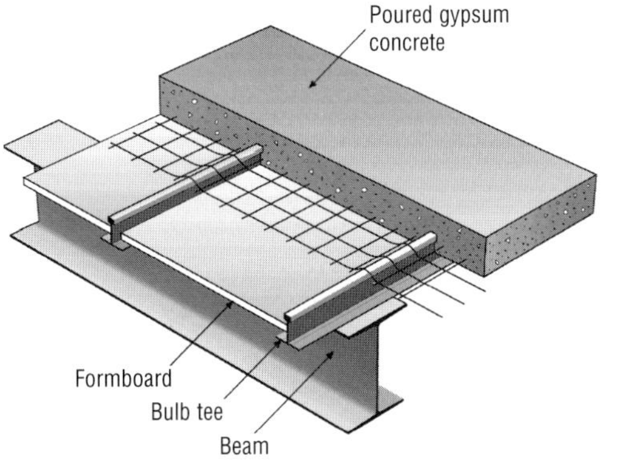

FIGURE 9
Cementitious Wood-fiber Deck

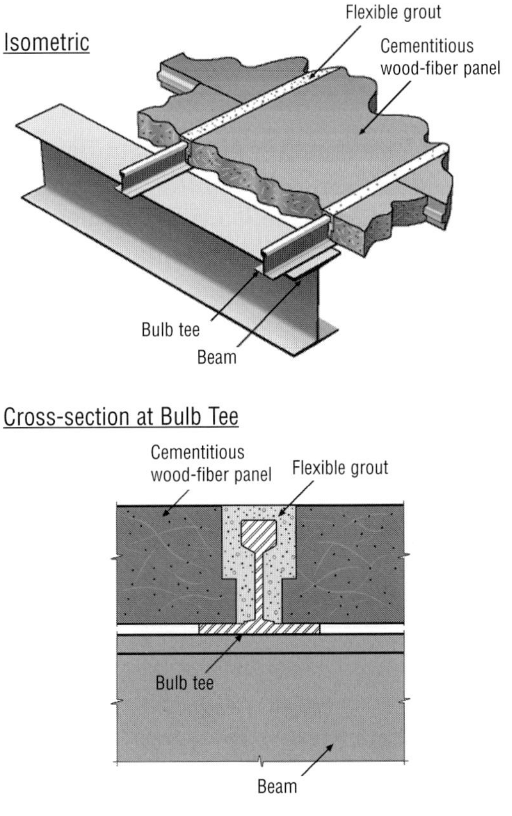

Gypsum Concrete

Gypsum concrete decks (figure 8) are made from gypsum, wood chips and water. It is mixed on-site and poured over wire mesh reinforcement supported by form boards. Generally made of fiberglass or gypsum, these boards are supported by bulb tees placed on the structural framing system. The slab is a minimum of 2 inches [50.80 mm] thick. The thickness of the slab can be varied to provide positive drainage.

Since gypsum decks are mixed on-site and poured in a wet condition, it is important that drying occurs downward to the space below the deck. The form boards, therefore, must be permeable. Painting of the form boards or installation of a suspended ceiling must be delayed until the deck is dry.

Cementitious Wood Fiber

Cementitious wood-fiber decks combine wood fibers with a cement or phenolic binder. The planks are generally 2-1/2 to 3 inches [63.50 to 76.20 mm] thick. They are available with either a felt covering, a cementitious topping or laminated foam insulation. This type of deck offers a combination of thermal insulation, acoustic properties and a finished ceiling.

Panel edges are either tongue-and-groove or routed. The tongue-and-groove panels are clipped to the structural steel framing system or nailed to wood beams. Those with routed edges are installed in bulb tees with grouted joints (figure 9). Resistance to wind uplift is provided by the clips, fasteners or grout, depending on the system.

To lessen the effects of panel joint movement on the roof membrane, insulation should be installed in two layers with joints offset. Insulation boards can be mopped to a nailed base sheet or mechanically fastened to the deck.

As with lightweight insulating decks, cementitious panels are unstable and can deteriorate in environments of high RH.

Combustible, Nailable Decks

Wood and Plywood

Wood is a traditional roof sheathing material. Its low weight, economy, ease of fabrication and availability make it a popular building material, particularly in the western United States. Wood is a combustible material, but it can be treated or a sprinkler system can be installed so that it can achieve the necessary insurance ratings. Wood has many advantages over those deck materials that can cause blisters, crack, provide inadequate fastener retention or corrode.

Although boards and planks (figure 10) are still used, most wood decks are made of structural wood panels. Particle board is not approved by roofing manufacturers as a suitable substrate for BUR. Structural wood panel systems can be designed to resist seismic loading. Industry standards for structural wood panels are published by APA—The Engineered Wood Association (APA), formerly known as The American Plywood Association.

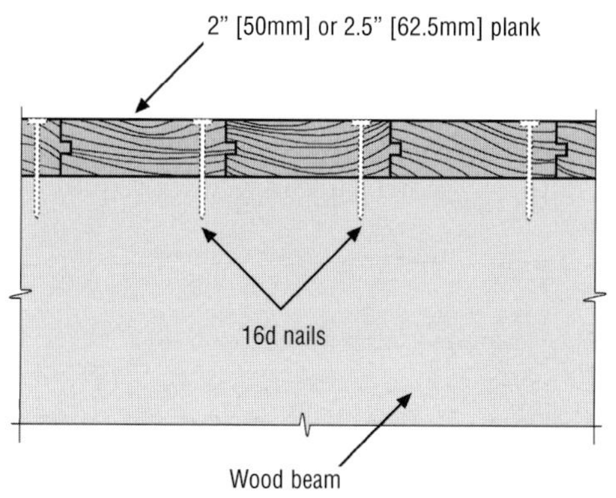

FIGURE 10
Wood Plank Deck

Structural wood panels should be either Exterior Grade or Exposure 1 and not less than 1/2 inch thick nominal. UL accepts 15/32 inch [11.90 mm] actual.

For structural wood panels, base sheets are secured to the deck with mechanical fasteners. Rigid insulation, if used, is mopped to the base sheet and not fastened to the deck. Batt insulation with an air space can be installed under the deck.

SUMMARY

- Concrete and prestressed concrete decks offer fire resistance, the ability to slope to drain and an attractive interior appearance. They are heavy and must be smooth enough to receive the roofing system, which is adhered with hot bitumen.
- Lightweight insulating concrete decks in their various forms offer lower dead loads, the ability to slope to drain and some thermal resistance. They must be designed for downward drying and should not be specified above environments of high RH. They are more suitable for warm climates, which do not experience significant freeze-thaw cycles.

- Steel's lightweight, ease and rapidity of installation, low cost, fire resistance and ability to accept fasteners make it the preferred deck nationwide. It must be covered with insulation to provide thermal resistance and a flat surface to receive a membrane.
- Wood is a traditional roof deck, but its lack of fire resistance restricts its use.

DRAINAGE

As a general rule, a fair roof with good drainage will outlast a good one with poor drainage. Proper drainage is important to the watertightness and longevity of a roof. Small holes or cracks in a membrane may be able to resist water infiltration because of the water's surface tension. However, if water attains any meaningful depth, hydrostatic pressure will force water through them.

Inadequate drainage can deteriorate a membrane, increase the potential for leaking and lead to organic growth. Roots from organic growth can penetrate the membrane at laps. When the plant dies in winter, the roots become hollow conduits that carry water into the roofing assembly. Ponding increases the strain on the roof and can lead to collapse if the roof deck is not capable of sustaining the load.

Positive drainage is often defined as the absence of ponding water 48 hours after the last rain. Most roofing manufacturers will not provide an owner a guarantee for a roof unless it exhibits positive drainage. NRCA and the Asphalt Roofing Manufacturers Association (ARMA), among others, advocate sloping roofs to drain. The three model building codes require that roofs be sloped a minimum of 2 percent. An architect who neglects to provide positive drainage does so at great risk.

FIGURE 11
Sloping Structure

Although a slope of 1 percent will provide positive drainage, a minimum slope of 2 percent is recommended on new buildings to accommodate settlement, creep and tolerances. Long span wood beams and lightweight structural concrete members, for example, can creep so much that a level roof can deflect after a few years, creating an unanticipated pond.

FIGURE 12
Wet Fill Applied to Roof Deck

Positive drainage can be achieved by any of the following:

- the structure (figure 11)
- wet fills (figure 12)
- tapered insulation (figure 13).

FIGURE 13
Tapered Insulation on Roof Deck

Drainage can be obtained by sloping the girders, beams and joists that support the roof deck. Cast-in-place concrete is easily contoured to create slopes on all kinds of roof configurations. On gypsum decks, the purlins can be sloped or the thickness of the gypsum can be varied.

Wet fills are composed of foamed cement or cement with lightweight aggregates, such as vermiculite or perlite. They can be easily formed into slopes. Wet fills must be designed to dry downward because moisture cannot escape through BUR, even if vents are provided in the roof assembly.

One manufacturer recommends a 2-inch-thick [50.80 mm] fill with stepped foam insulation boards over concrete or existing BUR roofs. This system is more successful on roofs in southern climates, where entrained moisture can dry quickly, than in environments where frequent freezing occurs. Wet fills should not be installed over metal decks unless the decks are galvanized and slotted to provide downward drying.

Factory-tapered insulation cut from flat board stock is commonly used to create slopes on flat roof decks. Virtually all types of insulation available on the market today can be purchased with various tapers. Combinations of flat and tapered boards provide great flexibility for contouring the roof. Layouts with peaks and valleys vary from two-way to four-way slopes. Crickets (figure 14) are often used to improve drainage in valleys, at eaves and along parapets.

FIGURE 14
Crickets Installed in Valleys Between Drains

DRAIN LOCATIONS

To drain low-slope roofs, internal drains located in valleys or at low points are preferred over scuppers or gutters. In climates subject to freezing, ice will usually form in scuppers or gutters. This can result in long, heavy icicles hanging from these external drains.

Drains should be located as close as possible to the midspan of a structural bay, where the deflection will be the greatest. The common practice of locating drains close to columns almost ensures that they will be high and dry. The maximum spacing between drains should be 75 feet [22.86 m]. The drain's leader should be offset from its tail piece by at least 24 inches [609.60 mm] to permit the drain to flex as the deck deflects, the building settles or the leader expands. The offset should be approximately 90 degrees, not 45 degrees as is frequently specified, to permit rotation around the riser. Alternately, a flexible rubber bellows can be installed between the drain and the leader. Bellows should always be used with PVC piping systems because they can better accommodate the greater movement of this material.

DRAIN TYPES

Internal drains are typically formed from cast iron (figure 15). The drain body consists of a bowl with flanges along the perimeter, lugs to receive bolts for a clamping ring and underdeck clamp, the clamping ring and a strainer. The underdeck clamp is critical to prevent the drain body from shifting and breaking the flashing seal.

The flanges of cast iron drains are thick and sloping. When a drain body is set directly on a deck, its flanges, in concert

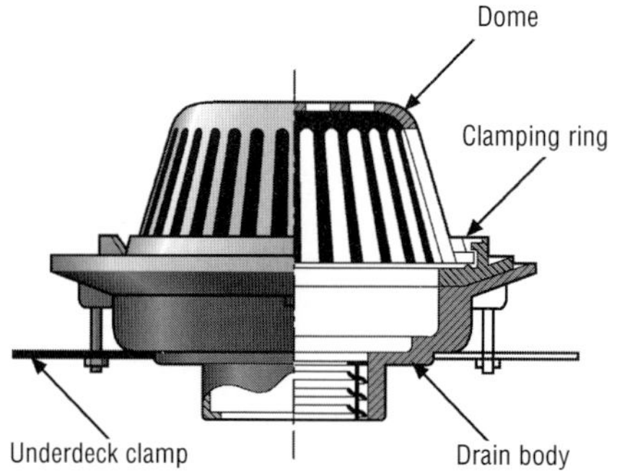

FIGURE 15
Cast Iron Roof Drain

with lead flashing and stripping, form a dam separating water on the roof from the drain bowl. A drain body, therefore, should be installed in a sump, or depression in the roof, at least 36 inches [914.40 mm] square so that it is below the adjacent roof level. Sumps are formed by sloping insulation. Because the insulation thickness at sumps is reduced, heat loss from the building will help keep the drains free of snow and ice.

Plastic drains (figure 16) are usually fabricated from PVC or ABS. Some plastic drains will soften from the high heat of the torch used to install flashing, so the architect should check with the drain manufacturer before specifying. Where additional above-deck insulation is not required, flat sheet-metal or plastic drains should be used.

FIGURE 16
Plastic Roof Drain

Some local codes require that roof drainage be restricted to avoid overloading sewer systems. This may require that drains be equipped with controlled-flow weirs, which retain a certain quantity of water on the roof for short periods of time. The weirs are adjustable and have an open top so that they will provide unrestricted flow into the storm drainage system if more than three inches [76.20 mm] of water backs up on the roof. Roofs for which controlled-flow drainage is planned must be structurally designed to support and retain the rainwater load during the prolonged drainage period.

OVERFLOWS

The model building codes require all roofs to be designed to support the weight of the water if all roof drains are clogged. If curb or parapet heights will increase the storm water load beyond the designed live load, overflow scuppers or drains (figure 17) must be provided. Some local codes dictate how high above the roof one should locate the top of the overflow drain or the bottom of the scupper. If not mandated, two inches [50.80 mm] is a safe rule of thumb.

FIGURE 17
Roof Drain (left) and Overflow Drain (right)

Check the local code before deciding on whether to install an overflow drain or a scupper. Some codes require that the overflows be connected to an independent piping system. This ensures that water will be discharged from the roof in the event that the leaders from the primary roof drains are blocked. This can be both expensive and difficult to install. One alternative is to take the discharge pipe from the overflow directly through an outside wall. The out flow from the piping can be directly below the roofline since water will only be flowing in an extreme condition. Visible overflow will serve as an excellent indicator of problems with the roof drains.

Overflow drains should not have a strainer unless required by local codes. Be sure that the overflow is clamped to the deck in the same way as the drain. Some plastic drains are manufactured with integral overflows that can save the cost of a separate pipe and flashing system.

One last word on drains: During roofing, drains must be plugged to avoid clogging them with debris. Remove plugs each night. Plugs should be equipped with telltale flags to assure their removal.

ROOF INSULATION

TYPES OF INSULATION

Insulation for low-slope roofs takes many forms. Rigid insulation boards or wet, lightweight-insulating fills can be installed above the deck and below the membrane. Batts or blankets can be hung below the deck. Cementitious panels comprising the deck itself can provide insulation. Insulation can also be placed above the roof membrane to form a protected membrane roof (PMR). A PMR is sometimes referred to as an Insulated Roof Membrane Assembly (IRMA), or an inverted or upside down roof. The scope of this monograph will limit the discussion to rigid insulation boards that are installed above the deck and below the membrane because they are the most common.

Rigid insulation materials include wood fiber; perlite; foamed glass and glass fiber; and expanded and extruded plastic foams (figures 18a–d). All are available in tapered form, which provides a slope for drainage on dead flat decks. Some are marketed as composites.

POLYSTYRENE FOAM	
DESCRIPTION	
Expanded (EPS)	Molded, fused beads, may be faced. K = 0.26 KSI = .037
Extruded (EEPS)	Fine cells; usually unfaced. K = 0.20 KSI = .028
STANDARDS	This product should be manufactured to meet the requirements of either ASTM C-578 or CAN/CGSB 51.20M.
CHARACTERISTICS	
Combustible	However, may be acceptable in some FM assemblies; especially when overlaid with single-ply membranes.
Attacked by solvents	Not suitable for direct adhesion with solvent-based adhesives.
Attacked by hot bitumen	Will melt. Can be adhered in hot bitumen if bitumen is allowed to cool to 200-250 F [93-121 C].
Dimensionally stable	Relatively high coefficient of thermal expansion but unaffected by water.
	EPS may shrink if not aged adequately at temperatures approaching 160 F [71 C].
Good handleability	Resistant to breakage. Will recover if compressed moderately.
Compressible	Should not be directly mechanically fastened without overlay. Compression varies with density.
Lightweight	EPS typically 1 to 1.25 pcf [16 to 20 kg/m^3] EEPS typically 2 pcf [32 kg/m^3]
Freeze-thaw resistance	EEPS is outstanding. Suitable for inverted roof membrane systems (protected membrane roofs).
Water resistant	EPS is very versatile. It is frequently embedded into lightweight concrete to improve R value, reduce weight.

FIGURE 18a

Typical Materials For Rigid Insulation Boards

Virtually all preformed thermal insulations are available as tapered boards. Manufacturer may provide shop drawings for layout of system.

SOURCE: RIEI

FIGURE 18b
Typical Materials For Rigid Insulation Boards
SOURCE: RIEI

WOOD FIBERBOARD	
DESCRIPTION	Consists of wood or sugar cane fiber, impregnated and coated with asphalt. K = 0.36 KSI = .052
STANDARDS	This product should be manufactured to meet the requirements of either ASTM C-208 or CAN/CSA A-247M, Type I (Roof Board).
CHARACTERISTICS	
Combustible	Can be FM approved over steel when it contains less than 4 percent asphalt and is nailed, not mopped. SMOLDERS: AVOID IN TORCH-APPLIED SYSTEMS.
Absorptive	Should be top-sealed for mopped or adhered systems.
Dimensionally unstable	Must be kept dry. Limit boards to 4 feet [1.2m] max dimensions.
Strong	Easily handled. Good peel resistance for adhered single-ply systems.
Incompressible	Good overlay for soft substrates, distributes stress of fasteners over EPS.
Heat resistant	Can be mopped with hot bitumen, used as overlay for plastic foams.
Solvent resistant	Can be adhered with common solvent-based adhesives.

GLASS FIBERBOARD	
DESCRIPTION	Consists of glass fibers bonded with a thermoset resin (phenolic); surfaced with asphalt-kraft which wraps around two edges. K = 0.24 KSI = .034
STANDARDS	This product should be manufactured to meet the requirements of either ASTM C-726 or CAN/CGSB 51.31M.
CHARACTERISTICS	
Low combustibility	Many FM/UL rated constructions.
Low absorption	Facer holds adhesives out of board.
Dimensionally stable	Resists warping.
Easily handled	Resists abuse.
Compressible	More compressible than other products. May require overlay or taped joints when thick boards are used, heavy roof traffic expected. When through fastened, fastener heads should recess into discs or lock to discs to avoid puncture through back-out.
Conformable	Can accommodate minor substrate irregularities such as in recover directly over old roof.
Heat & solvent resistant	

FIGURE 18c
Typical Materials For Rigid Insulation Boards
SOURCE: RIEI

PERLITE - FIBER	
DESCRIPTION	Consists of expanded beads of volcanic ore (perlite), wood fibers, asphalt and other binders. Usually top sealed with mineral-filled asphalt coating. K = 0.36 KSI = .052
STANDARDS	This product should be manufactured to meet the requirements of ASTM C-728.
CHARACTERISTICS	
Low combustibility	Many FM/UL constructions. Used as fire protection for other products.
Low absorption	Sealed surface holds bitumen out. Will absorb water if soaked.
Dimensionally stable	Some tendency to cup if prolonged wetting takes place. Better than wood fiber.
Acceptable handling	Boards can break if roughly handled. Easy to shape and cut.
Good compression resistance	Accepts fastener load, roof traffic.
Less conformable	Needs smooth substrate.
Heat & solvent resistant	

FOAMED GLASS	
DESCRIPTION	Cellular glass blocks or kraft-faced boards. K = 0.35 KSI = .050
STANDARDS	This product should be manufactured to meet the requirements of ASTM C-552.
CHARACTERISTICS	
Incombustible	
Low absorption	
Dimensionally stable	
More difficult handling	Tends to pulverize or shatter if roughly handled.
Objectionable odor when crushed	
Very high compression resistance	
Not conformable	
Heat & solvent resistant	

FIGURE 18d
Typical Materials For Rigid Insulation Boards
SOURCE: RIEI

POLYURETHANE & POLYISOCYANURATE (ISOBOARDS)	
DESCRIPTION	Cellular foams with facers of glass fiber, foil or roofing felt. K = 0.16 KSI = .023 Isocyanurates may be considered to be more heat and fire retardant members of the urethane family.
STANDARDS	Unfaced products should be manufactured to meet the requirements of ASTM C-591 or CGSB 51GP21M. Faced products should be manufactured to meet the requirements of ASTM C-1013 or CGSB 51.26M.
CHARACTERISTICS Low combustibility	Isoboards may be directly applied to steel roof decks and meet FM/UL.
Non-absorptive	Facers hold adhesives.
Dimensional stability	May vary. Some products will warp when wetted or if corners of boards are not well anchored. Relatively high coefficient of thermal expansion.
Affected by hot bitumen	May "bubble" when hot-mopped. NRCA Bullentin #9 recommends overlay for hot BUR.
Unstable "R" value	Cell gas is gradually diluted by diffusion of air into cells. Use "aged" R-values, following TIMA procedure (6 months-free air circulation).
Chlorofluorocarbons	Have used ozone-depleting blowing agents. Substitutions being evaluated.
Heat & solvent resistant	May be installed using either hot bitumen or common solvents.
Good handleability	Facers add to puncture resistance.
Lightweight	Typically 2 pcf [32 kg/m^3].

COMPOSITE INSULATIONS	
DESCRIPTION	Combinations of two or more products (from above) to provide synergistic effects.
Perlite Base with Foam	Perlite provides fire resistance on steel decks. Foam provides lightweight and good R-value. Should meet ASTM C-984.
Wood Fiber with Foam	Provides strong base to receive fasteners and adhered single-ply systems. Should meet ASTM C-1050.
Kraft Facers with EPS	Provides separator and slip sheet for PVC single-ply systems.

INSULATION BENEFITS AND CONCERNS

In general, there are several reasons for using insulation. Most states have energy codes that dictate overall thermal values for buildings. Insulation helps to meet these values by reducing the amount of energy that is needed to heat and cool a building. By such conservation of energy, the building owner will save money while the building occupants will be more comfortable. When properly installed, insulation in cold climates prevents condensation from occurring on a building's interior surfaces, which can be erroneously identified as leaks.

Rigid roof insulation provides additional benefits as well. It serves as a surface for the attachment of the roof membrane. It stabilizes the roof deck and deck components by reducing dimensional changes due to temperature fluctuations. And it provides a way of transferring stresses in the roof membrane to the structural system.

Rigid roof insulation does not come without risks. If improperly installed, it can increase the probability of condensation. Also, insulation will store heat energy, resulting in higher temperatures at the membrane that can accelerate the rate of membrane deterioration. Roof membranes over insulation experience higher temperatures in summer and lower temperatures in winter. It is not unusual for surface temperatures to reach 180 F [82.20 C] under summer skies and -50 F [-45.5 C] under cold clear nights. In addition, a sudden shower in the summer could reduce surface temperatures by 100 F [37.70 C] in a very short time. Temperature swings of those magnitudes can result in relatively large changes in the dimensions of the roof membrane, thus having an impact on the performance of the roof assembly.

SOME BASIC PHYSICS

Architects generally retain mechanical engineers to design heating and cooling systems and to offer advice on insulation requirements. An understanding of some basic building physics, however, is necessary to appreciate the function of insulating materials.

Heat Transfer

Heat energy is always moving from warm areas to cooler areas. Cold is never gained; heat is always lost or gained. Heat is transferred by three methods:

Conduction: the transfer of heat energy by direct particle to particle contact. If one end of a fireplace poker is placed in a fire, the other end of the poker will get hot after a period of time even though it is not actually in the fire. The heat is transferred by conduction through the particles of the poker material.

Convection: the transfer of heat energy by the movement of a liquid or gas. Natural convection occurs in a closed container when air is heated, causing it to expand and rise. The heated air displaces cooler air, which is forced down. In doing so, the air transfers heat energy from the lower part of the container to the upper part. This can happen in cavity walls, in attics and in spaces above dropped ceilings. As the column of air rises it rubs against the boundaries of the container, thereby transferring some of its heat to the container walls by conduction. Reducing convective currents in a building system will reduce heat transfer through the system.

Radiation: the transfer of heat energy by electromagnetic waves through air or a vacuum. Warm bodies radiate heat to cooler bodies or surfaces. The cooler element absorbs the radiation and becomes warmer while the space between the two bodies is not affected, remaining essentially at the same temperature. Air spaces lined with reflective materials can be effective insulators by reflecting radiant energy back to the building. Reflective materials are generally excellent conductors and, therefore, cannot provide any thermal reduction without an associated air space.

Moisture Flow

Water vapor exerts its own pressure on a building enclosure. If not properly accommodated, it can have a negative effect on roof performance.

Warm air can hold more water vapor than cool air. Warm air, therefore, will have a higher vapor pressure. Moisture vapor naturally migrates from areas of higher vapor pressure to those of lower vapor pressure.

In cold climates, the air in most buildings will contain more moisture than the outdoor air. During the winter in temperate climates, including most of the United States, the moisture within a building will also be greater than the air outside. In both cases, moisture from the inside will be driven to the outside.

The reverse holds true for tropic or subtropic climates. Here, moisture vapor is more often driven into the building because, for most of the year, vapor pressures are higher on the outside of buildings than on the inside. Except in very cold climates, the interior of a cold storage or freezer building is always at a lower vapor pressure than that of the exterior, so moisture migration will be inwards in this particular building type.

Water vapor can infiltrate a roof assembly by either diffusion or air movement. In diffusion, a difference in vapor pressure forces moisture through a permeable membrane. In air movement, air and moisture flow through discontinuities in a membrane due to air pressure differences or convection. Under some circumstances water vapor may be moving in one direction due to vapor pressure differences and in the opposite direction because of air movement.

Moving air can carry large amounts of water vapor. The amount of moisture migration due to air movement often exceeds that due to vapor pressure differences. A vapor retarder, therefore, is a key element in the control of moisture movement. Its integrity is equally important: Even small tears or holes can bring problems.

Air is saturated when it is holding as much water vapor as it can. Fog is saturated air. Absolute humidity is the actual amount of moisture in the air, in terms of pounds, grams, grains, or other measurement. Relative humidity (RH), which is expressed as a percent, is the ratio of the actual amount of moisture in a volume of air to the total amount of moisture that the air could hold if saturated. If a volume of air contains one unit of moisture (pound, gram, grain, etc.) but could hold two units at saturation, its RH is 50 percent. When air is heated, it can hold more moisture so its RH will go down while its absolute humidity will stay the same. The dew point is the temperature at which a mixture of air and water vapor reaches 100 percent RH and is considered saturated.

For example, air at 75 F [23.80 C] with an RH of 50 percent contains 64 grains of water vapor per pound [0.45 kg] of dry air (a pint [.47 l] of water contains 7000 grains). If the air temperature is reduced to 55 F [12.70 C], the RH will be 100 percent. This is the dew-point temperature. Saturated air at 55 F [12.70 C] contains 64 grains of moisture per pound [0.45 kg] of dry air. If the air is further cooled to 20 F [-6.60 C], the RH will still be 100 percent but now there is more moisture in the air than it can hold because saturated air at 20 F can contain only 22 grains of water vapor. The excess moisture must come out of the air as condensation. It takes the form of liquid water if the temperature of the condensing surface is above 32 F [0.00 C], of ice or frost if the temperature is 32 F [0.00 C] or below.

Terminology

Some of the basic units and equations needed for heat calculations in building technology are defined below.

Units of Energy

- Within the inch-pound system of measurement, The British thermal unit (Btu) is the heat energy required to raise the temperature of one pound of water by one degree Fahrenheit. The therm, which is the energy unit commonly used by gas companies in the United States to meter and sell energy, is equal to 100,000 Btu. The energy unit commonly used by electric companies is the kilowatt-hour (kWh).
- The unit of energy in the International System of Units (SI) is the joule (J), which is equal to one watt-second.
- The following equations indicate the relationships between the inch-pound and SI units:

  1 Btu = 1055 J
  1 therm = 105.5 MJ (megajoules)
  1 kWh = 3.6 MJ (k-thermal conductivity)

Power

Thermal conductivity (k) is the heat energy transmitted by conduction through one square foot of one-inch thick homogeneous material in one hour when there is a difference of one degree Fahrenheit perpendicularly across the two surfaces of material. Insulation with a lower k value is more efficient than one with a higher k value.

- Then k must be specified in Btu inches per square foot hour degree Fahrenheit, or (Btu x in)/(ft^2 x h x F).
- When calculations are done according to SI, A is measured in square meters, t in meters, and θ in degrees Celsius. Then k must be specified in watts per meter Celsius, or W/(m x C). This gives Φ in watts (W).
- Use the following formula to convert Φ from inch-pound to SI units:

  1 Btu/h = 0.2931 W.
- Use the following formula to convert k from inch-pound to SI units:

  1 (Btu x in)/(ft^2 x h x F) = 0.1443 W/(m x C).

Thermal Conductance (C) is the rate of heat flow for a specific thickness of a single material or of non-homogeneous materials laminated together, taken across unit area:

$$C = k/t$$

- In inch-pounds, C will be measured in Btu/(ft^2 x h x F).
- In SI units, it is expressed as W/(m^2 x C).

Thermal Resistance (R) is the reciprocal of thermal conductance or conductivity:

$$R = 1/C \text{ or } R = 1/K$$

R is an index of a material's ability to resist heat flow. The greater the value of R, the more efficient the insulation.

Overall Coefficient of Thermal Transmittance (U) is the rate of heat flow, per unit area, for a complete construction, such as a wall or roof, including all assembly materials and associated air films and air spaces. It is calculated by taking the reciprocal of the sum of all the thermal resistances in the assembly:

$$U = 1/(R_1 + R_2 + R_3 + \text{etc})$$

- In inch-pound units, U is expressed as $Btu/(ft^2 \times h \times F)$.
- In SI units, it is expressed as $W/(m^2 \times C)$.

SELECTING INSULATION

Heat Loss Calculations

Numerous books and computer programs are commercially available to calculate thermal losses and gains. The American Society of Heating, Refrigerating and Air-Conditioning Engineers (ASHRAE) publishes *The ASHRAE Handbook: Fundamentals*, the definitive work on calculating heat losses and gains for building elements and determining the design temperatures for cities in the United States. NRCA publishes a more simplified guide, *The NRCA Roofing and Waterproofing Manual*. The calculations are relatively straightforward when working with simple buildings. They require a professional consultant, however, as the building becomes more complicated and variables of internal gains from manufacturing processes, lighting and equipment are considered.

In order to perform these calculations, the architect or the consultant should have available the following:

- ASHRAE Design Data for Selected Cities, which contains the winter and summer temperatures throughout the United States, and can be found in *The ASHRAE Handbook* and *The NRCA Roofing and Waterproofing Manual*.
- The required U-value for the roof. This may be dictated by a state energy code. Otherwise, it can be calculated by the formula provided by the Energy Policy Act of 1992, which is contained in ASHRAE Standard 90.11989, entitled *Energy Efficient Design of New Buildings (Except New Low Rise Residential Buildings)*.
- A table of resistance values for air spaces and air films and another table for the thermal properties of various building materials, all of which can be found in *The ASHRAE Handbook* and *The NRCA Roofing and Waterproofing Manual*.
- A section drawing of the roof assembly from the top of the membrane to the underside of the ceiling below.

To select the appropriate insulation, the U-value for the assembly must first be determined. The reciprocal of this U-value is then calculated to determine the desired R-value for the assembly. The components are listed with their respective R-values; the R-value of the insulation is defined as X. The required R-value of the insulation is determined by subtracting the components' R-values from the desired assembly R-value. Based on this required R-value and other desired characteristics (described in the following section), an insulation type is selected and the thickness is determined from the manufacturer's literature (figure 19).

Desired Characteristics

In reality, no one type of insulation on the market has all the desired properties. The designer must prioritize the ideal characteristics listed below for the specific requirements of a particular project.

An architect should consider the following properties when selecting a rigid board insulation material:

Dimensionally stable: The material should not change dimension under varying temperature and moisture conditions. Shrinking insulation can cause splits in the membrane. An insulation material that experiences nonuniform changes can detach from the membrane or deck.

Lightweight: The material should not contribute significantly to the dead load on the roof. The installer should be able to handle it easily.

Tough: The material must be able to resist rough handling during installation without being damaged.

Resistant to high temperatures: The material must be able to resist the very high temperatures (greater than 375 F [190.50 C]) associated with BUR installation.

Sample Calculation of Insulation R-value

1. Determine required U-value for assembly from ASHRAE 90.1 — $U_{assembly}$ = 0.054
2. Calculate required R-value for assembly — $R_{assembly}$ = 1/0.054=18.52
3. List all components and their respective R-values:

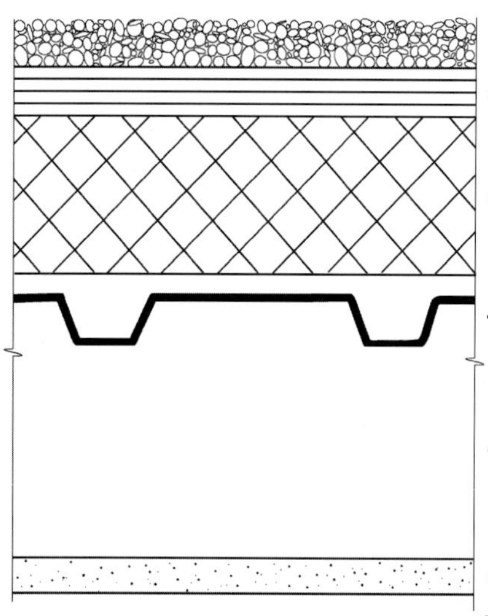

Component	R-value
Outside Air Film	0.17
BUR	0.33
Insulation	x
Metal Deck	0.00
Air Space	0.93
Ceiling Tile	1.56
Interior Air Film	0.61

4. Calculate total R-value for all components except insulation — R_{total} = ~~2.99~~ 3.60
5. Calculate required R-value for insulation ($x=R_{assembly} - R_{total}$) — 18.52 - ~~2.99~~ 3.60 = ~~15.53~~ =14.92
6. Select insulation type and thickness from manufacturer's literature:
   2.3 inches of isocyanurate insulation (R-value = 6.75/inch)

FIGURE 19
Example is for a building in Chicago.

Resistant to wetting: The material must be able to withstand occasional exposure to water without deteriorating or losing thermal resistance.

Compatible: The material must be compatible with the other roof materials. It must be able to bond to the deck or vapor retarder. It must not be vulnerable to chemical attack by other roof components or adhesives.

Attachment properties: The material must be able to be secured to the roof assembly with adhesives or mechanical fasteners to resist wind uplift.

Fire resistant: When combined with the other roof components, the material must meet all appropriate fire resistance requirements.

Shear strength: The material must have sufficient internal shear strength to transfer stresses to the supporting structural system without allowing the fasteners to pull through the insulation.

Low conductivity: The material must have a low k- or C-value to resist heat transfer efficiently.

Impact resistant: The material must be able to support the membrane above so that impact loads, such as hail or workmen, will not puncture the membrane or permanently crush the insulation.

Calculation of Cost Savings

The money saved by using insulation can be calculated by many different computer programs available on the market. Like heat loss calculations, cost savings calculations become more complicated as building occupancy and use become more complicated and should therefore be handled by a consultant.

FIGURE 20

If detailed appropriately, highlighted materials could serve as a roof vapor retarder since their perm ratings are less than 0.5.
SOURCE: RIEI

PERM RATINGS OF COMMON BUILDING MATERIALS

Material	Perm Rating (grains/hr•sq ft•inch Mercury)
BUR Membrane	0.0
Aluminum Foil 0.0	
Steel Deck	more than 1.0
Uncracked Concrete Structural Deck	about .05
1/4" [6.25 mm] Plywood with Exterior Glue	0.7
3/8" [9.38 mm] Gypsum Wall Board	50.0
4-mil Polyethylene	0.08
6-mil Polyethylene	0.06
4-mil Polyvinylchloride (PVC)	0.5
Kraft Paper Laminate	0.25
No. 15 Asphalt-Saturated Felt	1.0
No. 43 Asphalt Saturated and Coated Felt	0.3
Insulations	
1" [25 mm] Cellular Glass	about 0.0
1" Polyurethane	about 1.0
1" Extruded Polystyrene	about 1.2
1" Expanded Polystyrene	about 4.0

SPECIFYING VAPOR RETARDERS

Vapor retarders—which can be plastic films, laminates of kraft paper and asphalt, or single plies of coated roofing felts—are installed to slow the flow of moisture from a building's interior into the insulation during winter conditions. The rate at which water vapor flows through a material, measured in perms, is called permeance. The lower a material's perm rating, the less likely vapor will flow through it. A roof vapor retarder should have a perm rating of less than 0.5 (figure 20). The architect should work with the mechanical consultant to determine if a vapor retarder is needed for a particular project. The need will be determined by the environmental conditions on both the interior and exterior of the building.

On the roof, a vapor retarder is located between the deck and the rigid board insulation. During cold weather, water vapor rising from the heated spaces of a building will come in contact with the retarder before reaching the insulation.

The retarder must be properly installed without holes or breaks. A sufficient amount of insulation is required above the vapor retarder to ensure that the temperature of the vapor retarder is warmer than the dew point; otherwise condensation will occur on the vapor retarder surface. If the temperature profile of the building indicates that the temperature at the vapor retarder will be lower than the dew point, additional insulation must be installed or moisture levels within the building must be changed (figure 21).

Sample Calculation of Temperature at Vapor Retarder

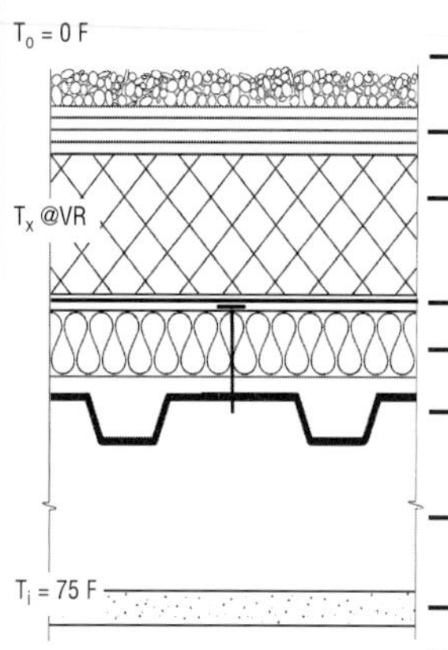

	RESISTANCES (R)	
Outside Air (f_o)	0.17	
Roof Membrane	0.33	
Roof Insulation	9.09	—
Vapor Retarder	0.12	0.12
Base Insulation	2.78	2.78
Steel Deck	0	0
Still Air	0.94	0.94
Ceiling Board	1.62	1.62
Inside Air Film (f_i)	0.61	0.61
	15.66 (ΣR)	6.07 (ΣR_x)

$$T_x=T_i-[(\frac{\Sigma R_x}{\Sigma R})(T_i-T_o)]$$

$$T_x=75\ F-[(\frac{6.07}{15.66})(75-0)]=75-(.3876)(75)=46$$

T_x=46 F at the Vapor Retarder Level

FIGURE 21

Using the calculated temperature at the vapor retarder from the example (T_x=46 F) and the dew-point temperature from a Psychrometric Chart (Tdp=55 F), determine whether the vapor retarder is warm enough to prevent condensation from occurring either in the insulation immediately below the vapor retarder (if insulation is installed at this location) or on the underside of the vapor retarder membrane. Because the temperature at the vapor retarder is below the dew-point temperature in this example, the vapor retarder is not warm enough to prevent condensation from occurring. Therefore, additional insulation must be added to the assembly above the vapor retarder (or insulation must be removed from below the vapor retarder).
SOURCE: The NRCA Roofing and Waterproofing Manual—Fourth Edition

ATTACHMENT OF INSULATION AND RETARDER

Attaching insulation or a vapor retarder to a roof deck with hot asphalt was the accepted method until 1953, when a devastating fire destroyed a huge General Motors plant in Detroit, Michigan. The extensive loss was blamed on the fuel provided by the asphalt that secured the insulation to the steel deck. Subsequently, lesser quantities of asphalt or cold adhesives were used in roofing projects, but these did not provide sufficient uplift resistance. In 1981, the industry shifted to mechanical fasteners, first around the perimeter of the building and then over the entire roof. Piercing type fasteners, applied by hammering, were replaced with a host of screw types (figure 22), many of which were designed for specific insulations.

FM lists the fasteners that they have tested and approved. It also publishes tables indicating the appropriate insulation and spacing for these fasteners. Fastener spacing will vary with design wind loads and board size, but is never less than one per two square feet.

FIGURE 22
Screw-type Fasteners

FIGURE 23
Nail One, Mop One Installation

On metal decks, the industry has adopted a two-layer system in which the first layer of insulation is mechanically fastened to the deck and the second layer is adhered with hot asphalt. This system, known as nail one, mop one (figure 23), offers many advantages, including the elimination of thermal bridging at the fasteners and the application of a vapor retarder above the bottom layer without piercing it.

MEMBRANE MATERIALS

A BUR membrane is made of three basic components: adhesive, reinforcement and surfacing.

FIGURE 24

Characteristics of Bitumen Roofing Materials

Characteristic	Asphalt	Coal-Tar Pitch
Cold Flow	No	Yes
Aging Method	Oxidizes	Sublimes
Preferred by Worker	Yes	No
General Availability	Yes	No
Self-priming	No	Yes
Results on Overheating		
with oxygen	Oxidizes	Loses volatiles
without oxygen	Softening point reduces (fallback)	
Deterioration in Water	Yes	No

ADHESIVES

An adhesive bonds a roofing system together and provides its waterproofing. Two types of adhesives are used: asphalt and coal-tar pitch (figure 24). Although similar in physical appearance and properties, these materials are not the same and, for the most part, are not compatible. Of the two, asphalt is more prevalent.

Asphalt

Asphalt is what is left at the bottom of the still when crude oil is processed. It is a very viscous liquid at room temperature. As it is heated, asphalt softens, becomes less viscous and turns into a free-flowing liquid. In its Standard D312, ASTM defines four types of asphalt (figure 25) according to the temperatures at which they begin to soften as determined by a standard test. The lower the soften-

FIGURE 25
SOURCE: ASTM

Physical Requirements of Asphalt in Roofing

	Type I		Type II		Type III		Type IV	
	Min.	Max.	Min.	Max.	Min.	Max.	Min.	Max
Softening Point, F [C]	135 [57]	151 [66]	158 [70]	176 [80]	185 [85]	205 [96]	210 [99]	255 [107]
Flash Point, F [C]	500 [260]	—	500 [260]	—	500 [260]	—	500 [260]	—
Penetration, units:								
at 32 F [0 C]	3	—	6	—	6	—	6	—
at 77 F [25 C]	18	60	18	40	15	35	12	25
at 115 F [46 C]	90	180	—	100	—	90	—	75
Ductility at 77 F [25 C], cm	10.0	—	3.0	—	2.5	—	1.5	—
Solubility in trichloroethylene, %	99	—	99	—	99	—	99	—

ing point the less viscous the material and the greater the tendency for it to flow. Asphalts from the oil refinery are modified at an asphalt plant to meet these standards. The most common way to modify asphalt is by blowing air through the material at high temperatures (about 500 F [260.00 C]). The longer the asphalt is subjected to the blowing process, the more viscous it becomes. The final blowing temperature is also critical, as will be explained below. Type I and II, known as softer asphalts, start to soften at about 135 F [57.20 C] and 150 F [65.50 C], respectively. The softer asphalts, applied at lower temperatures, are less likely to rupture when strained, more likely to self heal and better able to bridge moving joints. They are normally used on roofs with slopes of less than 4 percent. These roofs depend on their waterproofing qualities to maintain their integrity. The softer asphalts are also very important for pour, or flood, coats and for embedding surfacings.

Types III and IV, the harder asphalts, start to soften at 185 F [85.00 C] and 210 F [98.80 C], respectively. They are used on roofs with slopes greater than 4 percent, where resistance to slippage is paramount. They are also used in areas where adhesion of a material is critical, such as at the installation of base sheets, insulation, flashings and glass-fiber felts. In addition, the harder asphalts are commonly used in warm climates, where softer asphalts might result in slippage even on low slopes.

The ideal viscosity for roofing asphalt is one that will provide the optimum uniform waterproof film combined with sufficient adhesion for bonding the system together. The temperature needed to achieve these desired conditions is termed the equiviscous temperature (EVT). A BUR applicator attempts to maintain the EVT, plus or minus 25 F [-3.80 C], at the point where the asphalt is applied to the reinforcement felts. The EVT provides a viscosity rating of 50 to 200 centistokes (about the viscosity of 20w motor oil). If the bitumen film is too thin it may not provide a continuous waterproof barrier. Skips, or holidays, in the film will allow water to pass through the membrane, resulting in leaks. Film that is too thick will exhibit poor adhesive qualities that can result in slippage of the membrane on other than very low slopes.

Since heat is lost between the kettle or tanker, where the material is heated, and the point of application, the temperature at the kettle or tanker must be somewhat higher than the EVT. When air temperatures are cold, as in winter, more heat must be added to compensate for heat losses between the kettle or tanker and mopping point. If the material is heated too high, however, it will reach its flash point, the temperature at which vapors from the asphalt will ignite. Also, if the asphalt is heated above its final blowing temperature for more than 4 hours in a closed vessel, such as a kettle or tanker, it can lower the material's softening point. Known as a fallback, this will change the material's flow characteristics.

To achieve good performance, therefore, the applicator must know the following:

- asphalt type
- EVT needed at the point of application
- flash point of the asphalt
- final blowing temperature when asphalt was modified at the manufacturer's plant.

Coal-tar Pitch

Coal-tar pitch, described in ASTM D450, is the residue at the bottom of the container after coal is coked for the processing of steel. Coal-tar pitch has a low softening point (126 F [52.20 C]) and will self heal breaks or abrasions under normal summer roof conditions. However, it becomes a brittle solid at approximately 40 F [4.40 C] and, therefore, is more susceptible to physical damage from roof traffic in cold climates. Because of its ability to flow, coal-tar pitch should not be used on roof slopes greater than 2 percent.

Once the predominant adhesive in the roofing industry, coal-tar pitch now accounts for about 5 percent of the market. It is still specified in some areas, including Chicago, Pittsburgh and south Florida. Its decline has been attributed to changing market forces and health concerns. Three types (figure 26) are available: Type I is used for BUR. Type II is used primarily for waterproofing applications. And Type III, a low-fuming product, is a variation designed to eliminate the less desirable characteristics of coal tar while maintaining its desirable ones.

Incompatibility

Asphalt and coal tar react chemically with one another to form an oily interface with loss of adhesive bond. The coal tar hardens and cracks while the asphalt softens. The incompatibility is greatest when the two materials have similar softening points. The incompatibility problem is reduced when the asphalt has a high softening point (Type III or Type IV), is mixed with mineral fibers, or both. With high softening points, hard asphalts can be safely used as flashing material adjacent to coal-tar pitch.

FELT, MAT AND FABRIC REINFORCEMENTS

Non-woven felts and mats and woven fabrics give strength to a roof membrane in the same way that steel reinforcing bars do to concrete. Asphalt and coal-tar pitch have little tensile strength. The felt, mat or fabrics provide the strength needed to resist movement in the membrane due to temperature changes or building move-

FIGURE 26
SOURCE: ASTM

Physical Requirements of Coal-Tar Pitch for Built-Up Roofing

	Type I	Type II	Type III
Water max, %	0	0	0
Specific gravity, 25/25 C	1.22 to 1.34	1.22 to 1.34	1.22 to 1.34
Softening point (ring-and-ball), F [C]	126 to 140 [52 to 60]	106 to 126 [41 to 52]	133 to 147 [56 to 64]
Flash point, Cleveland open cup, min, F [C]	374 [190]	347 [175]	201 [205]
Total pitch soluble in carbon disulfide, %	72 to 85	72 to 85	72 to 85
Ash, max %	0.5	0.5	0.5
Total Distillate:			
32 to 572 F [0 to 300C], max, %	10	10	0
32 to 599 F [0 to 315C], max, %	—	—	0
32 to 680 F [0 to 360C] max, %	—	—	5
Specific gravity of distillate from 0 to 32 to 572 F [300 C], min, 38/15.5 C	1.03	1.03	—
Softening point (ring-and-ball) of residue from distillation to 572F [300 C], max, F[C]	176 [80]	176 [80]	—

ment. Reinforcements are made from a variety of materials using several manufacturing processes (figures 27 and 28).

Basic materials

Organic Felts

The earliest BUR systems relied on organic felts. Once made of rags, the felt is now manufactured from cellulose fibers on papermaking machines. The fibers are oriented in the direction of the roll, resulting in a product that is twice as strong in one direction (longitudinally) than the other. The material is fed through a saturator, which impregnates the cellulose fibers with very soft asphalt or coal-tar

Typical Organic Roofing Products

FELT MILL		SATURATOR			COATER		FINISHED PRODUCT				
Dry Wt. lb/100 ft²	Gage mils	Sat. %	Sat. Wt. lb/100 ft²	Prod. Wt. lb/100 ft²	Coating Wt. & Surf. lb/100 ft²	Product Wt. lb/100 ft²	Description	ASTM Number	Strength lb/in MD	Strength lb/in CMD	Ash Percent
5.2 Min.	27	120 Min.	6.2 Min.	11.5 Min.	—	—	No. 15 Asphalt Saturated Roofing Felt	D226	30 Min Type I	15 Min.	10 Max.
10 Min.	50	150 Min.	15 Min.	26 Min.	—	—	No. 30 Asphalt Saturated Roofing Felt	D226	40 Min. Type II	20 Min.	10 Max.
5.2 Min.	27	140 Min.	7.3 Min.	13 Min.	—	—	No. 15 Tar Saturated Roofing Felt	D227	30 Min.	15 Min.	—
5.2 Min.	25	140 Min.	7.2 Min.	—	18 Min.	37 Min.	No. 40 Asphalt Coated Base Sheet	D2626	35 Min.	20 Min.	—
10 Min.	50	160 Min.	16 Min.	—	18 Min.	54.6 Min.	No. 55 Asphalt Coated Roll Roofing	D224	45 Min. Type II	—	—
10 Min.	50	150 Min.	15 Min.	—	Granule Surfaced	74 Min.	No. 90 Asphalt Mineral Surfaced Roll Roofing	D249	—	—	10 Max.

FIGURE 27

SOURCE: RIEI

Typical Glass-Fiber Roofing Products

MAT MILL	COATER			FINISHED PRODUCT					
Dry Wt. lb/100 ft²	Coating Wt. lb/100 ft²	Surfacing Wt. lb/100 ft²	Product Wt. lb/100 ft²	Description	ASTM Number	Strength lb/in. MD	Strength lb/in. CMD	Moisture Percent	Ash Percent
1.5 Min.	6.3 Min.(unfilled)	2.2 Max.	8.4 Min.	Standard Ply Sheet Type III	D2178	22 Min.	22 Min.	1 Max.	70-88
1.7 Min.	3.0 Min.(unfilled)	3.2 Max.	6.0 Min.	Heavy-Duty Ply Sheet Type IV	D2178	44 Min.	44 Min.	1 Max.	70-88
1.9 Min.	3.0 Min.(unfilled)	6.0 Min.	6.0 Min.	Ply Sheet Type VI	D2178	60 Min.	60 Min.	1 Max.	70-88
1.7 Min.	3.0 Min.(coal tar)	3.0 Max.	6.0 Min.	Coal Tar Glass Felt Type I	D4990	44 Min	44 Min.	1 Max	70-88
				Premium Tar Glass	—	60 Min.	60 Min.	—	—
1.7 Min.	— (Max. 55% Filler)	24.0 Min. (Granules)	63.2 Min.	Mineral Surfaced Glass Fiber Cap Sheet	D-3909	—	—	1 Max	—
1.5 Min.	20.0 Min. (Max. 60% Filler)	8.0 Min. (Granules)	50 Min.	Asphalt Coated & Mineral Surfaced Glass Fiber	D4897 Type I	22 Min.	22 Min.	1 Max.	70-88
1.7 Min.	22.0 Min. (Max. 60% Filler)	8.0 Min. (Granules)	55 Min.	Venting Base Felt	Type II	44 Min.	44 Min.	1 Max.	70-88
1.4 Min.	5.6 (Asphalt)	—	12.8 Min.	Asphalt Coated Type I Glass Fiber	D4601	22 Min.	22 Min.	1 Max.	70-88
1.7 Min.	7.0 (Asphalt)	—	14.5 Min.	Base Sheet Type II	D4601	44 Min.	44 Min.	1 Max.	70-88

FIGURE 28

SOURCE: RIEI

pitch. The finished material is not fully saturated. ASTM Standards D226 (asphalt) and D227 (coal tar) provide criteria that must be met by the manufacturer. The finished felts that will be used as plies in a BUR system must be perforated by the manufacturer so that air trapped between the felts during installation can escape. Organic felts can absorb moisture and rot when exposed to wet environments. In comparison to other reinforcing options, they are weaker but more flexible. Organic felts easily conform to irregular surfaces.

Glass-fiber Mats and Fabrics

Glass-fiber mats are made of glass filaments bound in a thermosetting binder. They are very strong: tensile strengths exceed 60 pounds per inch (psi) [413.70 kilopascal (kPa)]. Since the mats are not absorbent, they are coated with asphalt. They have a naturally open texture that allows liquid bitumen, both asphalt and coal-tar pitch, to flow into and through the mat while letting air between plies escape during installation. It is not necessary, therefore, to perforate the mats. Because glass mats are not absorbent, they are not subject to rotting problems more commonly associated with organic felts. Most BUR membranes applied today use glass mats because of these advantages. However, glass-fiber mats do have some drawbacks. They are inherently stiff, so it is difficult to make them conform to irregular surfaces. They tend to want to return to their original shape. Known as memory, this tendency makes it difficult to use them for flashing. Hence, more flexible woven-glass fabrics that meet the criteria of ASTM D1668 are used for these particular applications.

Specially Coated and Surfaced Felts and Mats

All of the basic asphalt-treated felts and mats as they are described above may be used in BUR applications. They may also be coated during the manufacturing process with hard asphalts and stabilizers—such as silica, mica or talc—to make them more resistant to the environment. Mineral granules may be embedded in these coated sheets for added protection.

Organic felts and glass-fiber mats saturated with coal tar, however, cannot be coated or surfaced with granules because the low softening point of the coal tar will not provide a good adhesive for the granules.

Applications

Basic organic felts and glass-fiber mats are used as ply felts in multiple-ply BUR applications.

Specially coated organic felts and glass-fiber mats are used as base sheets in the membrane assembly. In this application, the coated sheets are usually punched with holes to enhance adhesion to the underlying surface and to allow air to escape from underneath the material during installation.

Specially coated organic felts without perforations may also be used as vapor retarders.

Mineral-surfaced sheets are used as cap sheets for surfacing and as the top ply in base flashings.

SURFACINGS

Surfacing serves many roles. All types provide shielding from solar radiation, which degrades the bituminous materials. Light-colored surfacing can reduce a roof's surface temperature, thereby slowing down the aging process of the bituminous materials. Aggregate surfacing can increase the membrane's resistance to wind uplift and to mechanical damage from foot traffic, hail stones or other such abuses. Coatings applied to the roof surface can enhance a membrane's resistance to fire. Although some manufacturers specify unsurfaced BUR systems in their catalogs, such membranes are not recommended for anything but temporary structures.

Surfacings are subject to weathering and cracking due to temperature changes and the passage of time. The surfacing may thin in areas due to erosion from wind or the flow of bitumen during periods of high temperatures. The surfacing also can be damaged by roof traffic and falling objects. Surfacings, therefore, do not necessarily contribute to the waterproofness of the system.

Aggregate Surfacing

Aggregate surfacing is the most common type of covering associated with BUR. It consists of an aggregate embedded into a flood coat of bituminous material. The aggregate can be waterworn, or bankrun, gravel; crushed stone; or slag. Other materials, such as marble and brick, have also been used but must be carefully selected: Those that are translucent, brittle or have excessive amounts of dust or salts should not be used as BUR surfacings. Translucent surfacings cannot prevent the transmission of solar radiation. Brittle materials and those with excessive dust or salts will not properly embed into the flood coat and will erode away.

The aggregate must be of appropriate size and gradation to be properly embedded into the flood coat and to provide an unbroken surface of opaque material. ASTM Standard D1863 describes the sizes and distribution of aggregate needed to provide an acceptable surfacing. The most common gradations used are Size 6 and Size 67. Size 6 allows diameters from 3/4 to 3/8 inch [19.00 to 9.50 mm] while Size 67 allows diameters from ¾ to 3/16 inch [19.00 to 4.75 mm]. Aggregates not meeting Size 6 or Size 67 criteria will not provide the needed protection, are more likely to erode when subjected to wind or flowing water, and will also be more difficult to embed in the flood coat.

Ideally, the surfacing material should be fully embedded into the flood coat. This, however, is not usually possible under field conditions. With careful application, high rates exceeding 50 percent are achievable.

A flood coat of asphalt should be applied at the rate of 60 pounds per square [27.20 kg per m^2] of roofing; that of coal-tar pitch at 75 pounds per square [34.00 kg per m^2]. A square equals 100 square feet [9.29 m^2]. To achieve these high rates, an applicator must pour hot bitumen from a dispenser instead of applying it with a mopping system. Laboratory and field data confirm that embedment does not occur when the flood coat is low.

Crushed stone and gravel are applied at a rate of about 400 pounds per square [181.40 kg per m^2] (figure 29). Slag is applied at about 300 pounds per square [136.00 kg per m^2].

FIGURE 29
Gravel Surfacing in Flood Coat of Asphalt

Double-pouring is a technique in which the surfacing is applied twice in two flood coats and two layers of aggregate. It can achieve higher rates of embedment. An opaque aggregate can be used in the first application to shield against solar radiation, a light-colored aggregate such as marble can be used in the second application to provide solar reflectance. Although the two-step application is heavier and more expensive up front, the energy saved from the higher reflectance of the roof surface coupled with the added protection for the roofing membrane might make it less expensive in the long run.

Aggregate surfacing cannot always be used: Some local building codes prohibit such surfacing on high-rises and in areas prone to very high winds. Loose aggregate is a potential hazard to jet engines so aggregate surfacing is also prohibited by some airports. In such cases a smooth-surfaced roof membrane must be installed instead.

Smooth Surfacings

A smooth-surfaced roof is formed when a basic membrane reinforced with inorganic felts is coated with hot Type IV-asphalt at a rate of about 15 to 25 pounds per 100 square feet [7.32 to 12.21 kg per 10 m^2] of surface. The hot asphalt can be substituted with cold-applied asphalt emulsions or asphalt cutbacks, which meet criteria specified by ASTM D2823, D2824 and D4479, that are applied at the rate of one gallon [3.79 litres (L)] of coating to 25 to 50 square feet [2.30 to 4.60 m^2] of roof surface.

Smooth surfaced systems have some advantages over their aggregate counterparts. They can be used on roofs that are too steep to support aggregate surfacing. If subjected to accumulation of debris from manufacturing processes, these roofs are easier to clean. And they allow much easier inspection of the membrane for physical damage or wear that can precipitate leaks or premature deterioration. This type of surfacing can also contain a heat reflective material such as aluminum flakes or white pigments that reduces surface temperature and may also enhance fire resistance.

During installation, applicators must make sure the coating does not become too thick, or else it will alligator, or crack, due to shrinkage (figure 30). Many coatings are emulsions of clay suspended in water. These materials cannot be applied when it is raining or when rain is expected. Exposure to water during this early period will prevent these coatings from curing; they will end up in the roof drains instead of staying on the roof. When aluminum coatings are specified, the underlying asphalt must be allowed to weather for at least 1 month to prevent it from bleeding through the coating. The building owner must be informed that the smooth-surfacing coatings have a relatively short lifespan (3 to 4 years) and, therefore, the roof must be recoated at regular intervals. Smooth-surface roofs should have a minimum slope of 2 percent to ensure rapid runoff of rainwater to the drains.

FIGURE 30
Alligator Coating

Cap Sheets

Mineral-surfaced cap sheets used as the final ply is another surfacing option. The cap sheet is used for surfacing only; it is not relied upon to provide watertightness or strength to the membrane. Various colors are available, including light or white granules that will reduce surface temperatures and cooling loads in some installations. The cap sheets are useful where aggregate surfacing is not acceptable but a smooth surfacing cannot offer sufficient resistance to damage. Roofs using cap sheets for surfacing should have a minimum slope of 4 percent.

FLASHING

INTRODUCTION

Leaking rarely occurs in the field of the roof. It mostly occurs at the terminations and penetrations of the membrane. Flashing is any system that prevents water from entering at these locations. It is, therefore, one of the most critical components of a roof assembly.

Flashing is required where the membrane terminates at the following locations on a roof:

- rising walls and parapets
- roof edges
- roof penetrations
- mechanical equipment
- expansion joint covers and area dividers.

Flashing details can be found in many reference books and manufacturers' catalogs. *The NRCA Roofing and Waterproofing Manual* shows how to detail the most frequently encountered types of terminations and penetrations. *Architectural Sheet Metal Manual,* published by the Sheet Metal and Air Conditioning Contractors' National Association (SMACNA), is an excellent resource for detailing metal work. FM's *Loss Prevention Data Sheets* show several ways to achieve wind-uplift resistance at roof edges. These typical details, however, will usually have to be modified to suit the actual conditions of a particular building. Drawing the details in section may not be sufficient. Intersections of flashings should be studied and drawn in isometric; they should not be left to the discretion of the mechanic in the field.

Many potential flashing problems can be eliminated by avoiding pitch pans, separating curbed openings by at least 12 inches [304.80 mm], and keeping clerestory sills at least 12 inches [304.80 mm] above the roof membrane. For general rules regarding flashing design, see *figure 31*.

RISING WALLS AND PARAPETS

Roofs enclosed by rising walls or parapets are like trays that are required to hold water for short periods of time. The waterproof component of the roof assembly, the membrane, must be turned up on these walls to complete the waterproof container. The turned-up plies of the assembly are called the base flashing. They are applied in an operation separate from the application of the membrane plies. The top of the base flashing, which is susceptible to water penetration from rain running down the wall or parapet, must be sealed and protected with a cover that projects over the base flashing and sheds water away from it. This is the counterflashing, or cap flashing.

The top of the base flashing should be 8 to 10 inches [203.20 to 254.00 mm]

FIGURE 31

Flashings should be waterproof, durable, flexible, stable and compatible with the roof membrane.

General Rules for Flashing Design

Rule 1	Isolate roofing system when deck is not supported by perimeter walls.
Reason 1	Avoids putting stress on flashing that spans between the roof and wall.
Rule 2	Insure that water will immediately drain away from all junctures and penetrations. Use tapered insulation to direct flow of water away from penetrations.
Reason 2	Minimizes infiltration of water if flashing fails in these areas.
Rule 3	If possible, do not bury metal in the roofing membrane or flashings.
Reason 3	Avoids differential movement between flashing and membrane.
Rule 4	Provide pressure-treated wood nailers under all metal flanges at least as wide as the flange. Do not use fire-treated lumber.
Reason 4	Metal must be mechanically attached so that metal edges cannot cut the membrane. Fire-treated lumber can be corrosive if wet.
Rule 5	The top of the base flashing should be at least 8 inches [203.20 mm] above the roof membrane but no more than 12 inches [304.80 mm].
Reason 5	Prevents water from flowing over the top of the flashing and avoids flashing slippage.
Rule 6	Avoid pitch pans.
Reason 6	Pitch pans are a maintenance problem and are excluded by warranty.
Rule 7	Recess drains in sumps below the roof level.
Reason 7	Minimizes damming from drain flashing stripping.
Rule 8	Elevate roof expansion joints a minimum of 8 inches [203.20 mm] above the membrane.
Reason 8	Minimizes water intrusion if joint cover fails.
Rule 9	Secure wood nailers to deck, not to rising wall or parapet.
Reason 9	Prevents membrane shrinkage from pulling nailer fasteners from wall.
Rule 10	Use two-piece cap-and-receiver counterflashing on walls and parapets.
Reason 10	Permits reroofing without distorting counterflashing metal.
Rule 11	Use continuous cleats on faces of metal edges, gravel stops, coping covers and curb covers.
Reason 11	Provides positive restraint for wind uplift.
Rule 12	Extend counterflashing a minimum of 4 inches [101.60 mm] over base flashings.
Reason 12	Protects base flashing.
Rule 13	Group conduits and pipes into one opening in roof where possible.
Reason 13	Minimizes roof openings.
Rule 14	Maintain at least 12 inches [304.80 mm] between penetrations and 24 inches [610 mm] between penetrations and toe of cant strip.
Reason 14	Provides space for base flashings.
Rule 15	Obtain manufacturer's approval of flashing details.
Reason 15	Ensures that flashings will be included in warrantee.
Rule 16	Reduce roof penetrations to a minimum.
Reason 16	Reduces possible leaks and simplifies flashing.

above the plane of the roof to prevent wind driven rain from reaching the top of the base flashing. The counterflashing should cover the base flashing by at least 4 inches [101.60 mm]. When pavers are used or abuse of the flashings is expected, the counterflashing can extend down over the cant to the roof membrane. If this is done, the bottom edge of the metal should be folded back to form a hem that will stiffen the counterflashing. In high wind areas, clips may be needed to secure the bottom of the counterflashing.

Base Flashing

Materials

Base flashings are built up of materials similar to those used for the membrane: one or two plies of organic felts or glass-woven fabrics laminated with hot asphalt or cold flashing cement. Glass felts are not used because it is difficult to make them conform to irregular surfaces and their "memory" can cause them to return to their original shape.

After they are installed, the plies are covered with a fibrated, aluminum asphalt coating; a ply of glass-reinforced, granule-surfaced cap sheet; or a torch applied modified bitumen sheet. The torch-applied modified bitumen sheet is preferred because of its durability and flexibility. In some situations, special care must be taken in torch-application of base flashings because of potential for starting a fire.

Metal base flashings are not specified for BUR since their greater thermally induced movements would rupture the membrane.

Base flashing materials are stiff and cannot be bent to a 90-degree angle without cracking. A triangular cant strip of pressure-treated wood, wood fiber or perlite is installed to break a 90-degree angle into two 45-degree transitions. Pressure-treated wood cants are the strongest and can provide bracing for wood curbs, if needed. Wood cants will not deteriorate if subjected to moisture. However, they tend to warp or twist and must be secured with mechanical fasteners. They are more expensive than the other types of cants. Perlite and wood-fiber cants can be installed in hot asphalt or cold mastics.

Application

Base flashings are installed as follows:

- Masonry or concrete walls are primed.
- The BUR plies are carried up to the top of the cant strip and cut off.
- One or more plies of base flashing plus a cap sheet are extended a minimum of 4 inches [101.60 mm] out over the membrane. These plies and cap sheet are also extended a minimum of 8 inches [203.20 mm] and a maximum of 12 inches [304.80 mm] above the membrane.
- The top of the flashing plies is secured to masonry with large-headed nails spaced about 8 inches [203.20 mm] apart or to concrete walls with flashing cement. Hot asphalt or flashing cement can be used on other

FIGURE 32
Base Flashing for Wall-supported Deck
SOURCE: NRCA

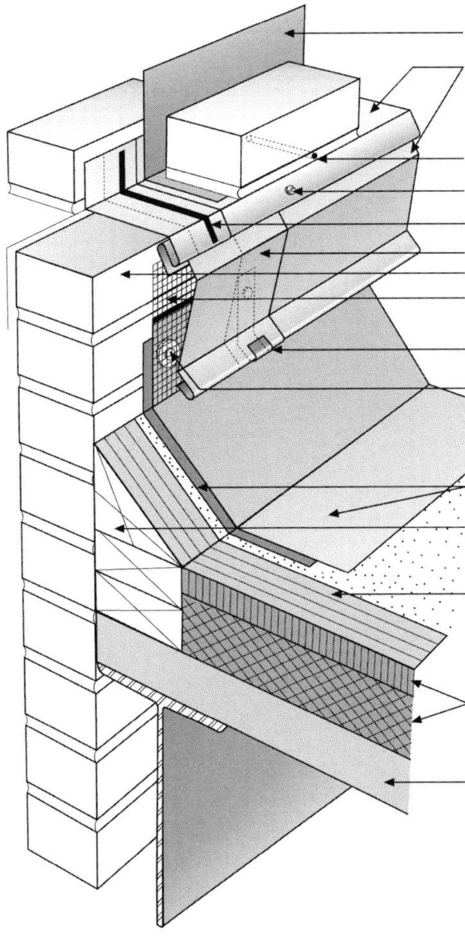

substrates. The top edge of the plies is sealed with flashing cement and fabric (figure 32).

- If the roof deck is supported independently of the walls and parapets, as in buildings that use tiltup wall construction, differential movement between the deck and walls can wrinkle or rupture the flashings (figure 33). To prevent this, the base flashing should be installed on a wood curb at least 8 inches [203.20 mm] high and separated from the wall by 1 or 2 inches [25.40 or 50.80 mm]. The space between the wall and curb should be filled with flexible insulation to reduce heat losses and to control condensation (figure 34).

FIGURE 33
Wrinkled Base Flashing
This was caused by differential movement between the deck and wall.

- Base flashing should not extend to the top of a masonry wall when the coping is more than 12 inches [304.80mm] above the roof deck. Masonry will invariably absorb water, usually through cracks between the mortar and masonry units. Covering one side of the wall with a moisture-impervious

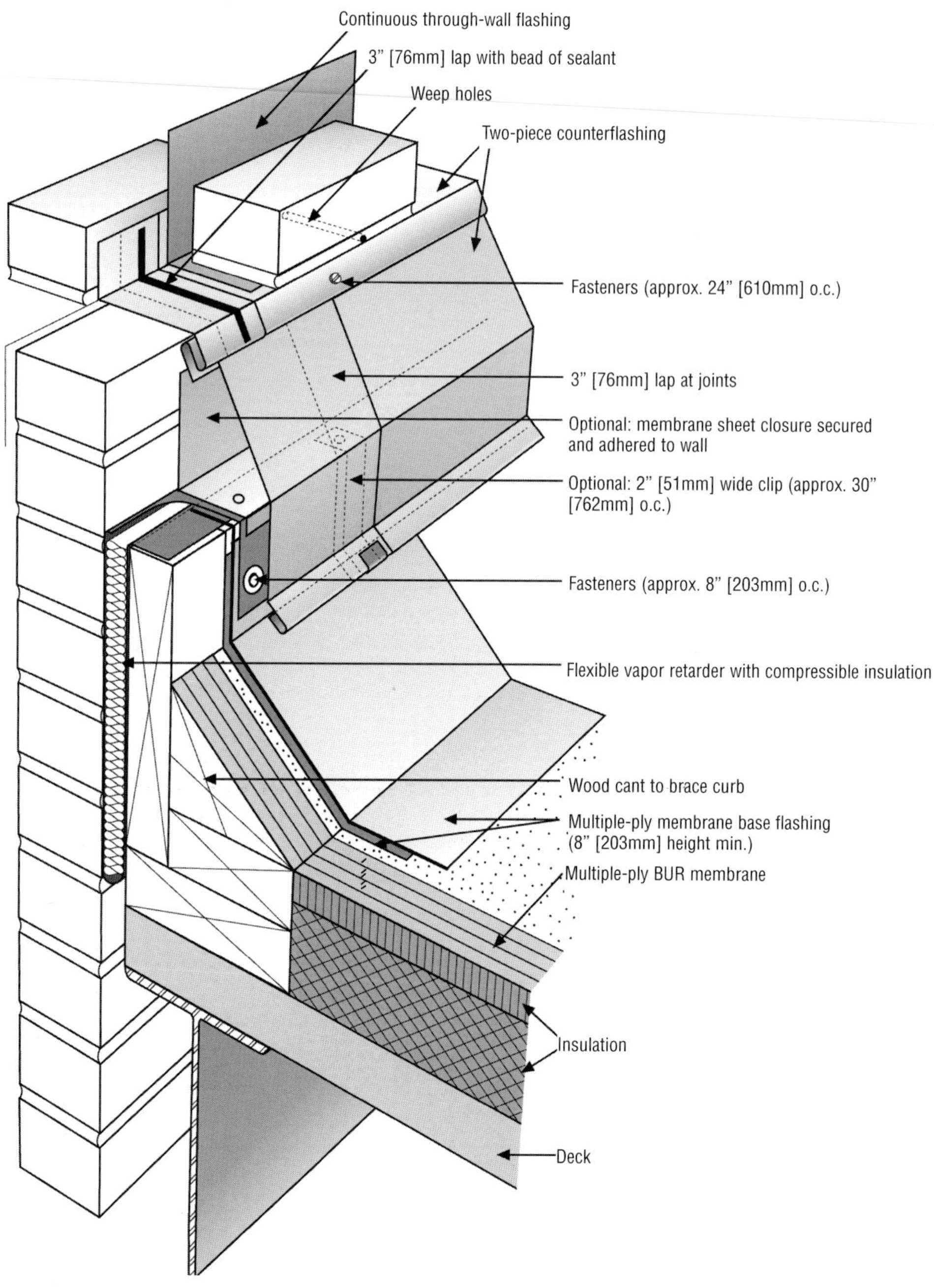

FIGURE 34

Base Flashing for Non-wall-supported Deck

SOURCE: National Roofing Contractors Association, The NRCA Roofing and Waterproofing Manual—Fourth Edition

material will inhibit the evaporation of this water. In cold climates the water will freeze, causing the masonry to spall. In addition, vapor pressures in the wall can cause the base flashings to blister or lose their bond to the walls.

Counterflashing

Counterflashings direct water away from the top edge of the base flashings. The counterflashing may be the exposed component of a throughwall flashing system; built into a masonry wall; inserted into a reglet; surface mounted on a concrete wall; or the vertical component of a two-piece cap and receiver.

FIGURE 35

Installation costs are given relative to stainless steel. For example, if stainless steel costs $100 per linear foot to install, aluminum installation costs $60 per linear foot. *SOURCE: The Building Envelope: Solutions to Problems Seminar Series Session 1—Exterior Wall Waterproofing,* Simpson Gumpertz & Heger Inc., Arlington, MA

Properties of Flashing Materials

Material	Advantages	Disadvantages	Installed Cost	Minimum Thickness
Stainless steel, 302, 304, 2D, dead soft, annealed	Hard, very durable, impervious	Difficult to solder, bend, and form	100%	0.012 inch [0.32 mm]
Cold-rolled copper	Impervious, flexible, durable, easily formed and joined	Damaged by excessive flexing; stains surfaces where water runs off	75%	16 ounces [453g]
Lead-coated copper	(See copper) Does not stain surfaces below; easy to paint	(See copper) Requires special care to solder	90%	16 ounces [453g]
Aluminum	Fairly durable; can be formed; corrosion-resistant except in the presence of lime	High thermal coefficient; cracks at fasteners and bends; can't be field-soldered; corroded by lime	60%	0.032 inch [.80 mm]
Galvanized steel	Hard, easily formed, impervious, low thermal coefficient, easy to paint	Subject to early corrosion in acidic and salty air; difficult to solder	80%	26-gauge [0.48 mm]

Materials

Counterflashings can be fabricated from many materials (figure 35). Galvanized steel should not be used near coastal regions or in heavily polluted areas because of its susceptibility to corrosion. Concealed flashings in masonry walls formed of modified bitumens, laminated metal foils, EPDM or PVC should not be extended beyond the face of the wall to form counterflashings. They may have poor UV resistance and are not stiff enough to provide adequate resistance to wind uplift. A metal receiver extending 4 inches [101.60 mm] into the wall and sealed to the concealed through-wall flashing should be used instead (figure 36).

FIGURE 36

Two-piece Receiver and Cap Flashing with Through-wall Flashing

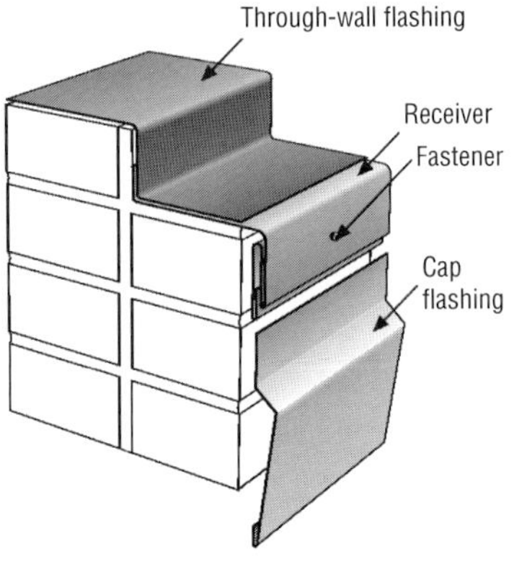

Application

Multiple wythe masonry walls and parapets should be flashed with through-wall flashing. The horizontal leg of the counterflashing should extend a full brick wythe deep and turn up under the through-wall flashing (figures 32 and 36). If the horizontal leg is less than a full brick wythe deep, water can flow down the interior of the wall and bypass the inner edge of the leg. This can appear to be a leak in the roof flashing when, in reality, it is a leak in the masonry wall. Single wythe masonry walls should be covered with a panel system that can provide an adequate counterflashing system (figure 37).

On concrete walls, surface-mounted counterflashings (figure 38) are just as effective as reglets. They are less expensive and easier to install on buildings that have many corners. Special precautions are necessary when detailing counterflashing systems for precast walls and tilt-up structures: The vertical joints between the panels can be a source of water entry into the roofing assembly and must be covered with a batten strip. Surface-mounted counterflashings should be sealed at their tops with a noncuring high solids butyl caulk. Elastomeric sealants such as silicones and urethane are not appropriate here since this is a nonmoving joint.

The roof membrane and flashing system will not last as long as the building, therefore reroofing can be expected several times during the life of the building. A

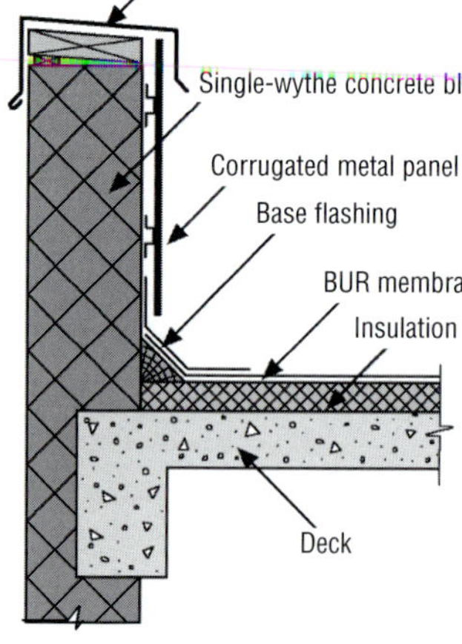

FIGURE 37
Panel System for Counterflashing Single-wythe Concrete Block Walls

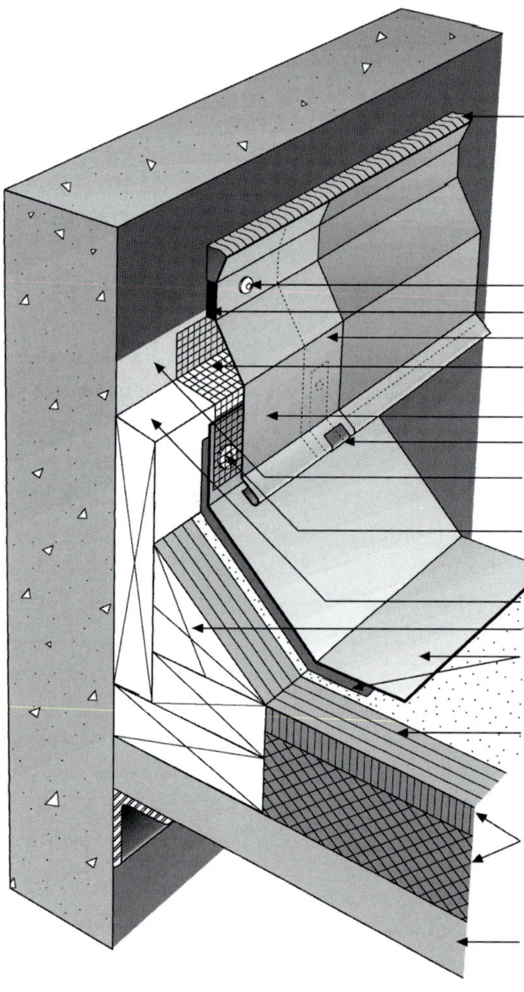

FIGURE 38
Surface-mount Counterflashing for Concrete Walls
SOURCE: NRCA

two-piece counterflashing, consisting of a cap and receiver (figure 36), allows the roofer to remove the exposed part of the flashing and install new base flashing without having to bend the counterflashing .

ROOF EDGES

Roof membranes that do not terminate at rising walls, parapets or expansion joints are terminated at metal edges. A formed metal edge is always used where a roof drains at the perimeter. It protects the end of the membrane from abuse and wind uplift as well as providing an architectural finish. On aggregate-surfaced roofs, the metal edge serves as a gravel stop. The use of metal roof edges embedded in a roof membrane, although described below, should be avoided if at all possible. It is better practice to use a raised roof edge.

Metal Edges

Materials

Metal edge and gravel stops are fabricated from the same materials as counterflashings. They may also be extruded from aluminum. Light metal gravel stops are generally fabricated from 0.024-inch-thick [.60 mm] steel, 16-ounce [453.6 grams] copper or 0.040-inch-thick [1.02 mm] aluminum. Extruded shapes are generally 0.060 inch thick. The horizontal leg, or flange, of the metal edge should never be less than 4 inches [101.60 mm] wide.

Application on a Roof Membrane

The light-gauge metal edge is primed, bedded in roof cement and secured through its horizontal flange to a treated wood nailer below the membrane. The nailer, which is anchored to the building, must be at least as wide as the flange. Two staggered rows of nails spaced 3 to 4 inches [76.20 to 101.60 mm] on center are used to secure the flange. The closely spaced nails minimize thermal movement in the roof edge caused by temperature changes. The flange is then protected with two plies of stripping felts that extend out onto the roof membrane (figure 39).

FIGURE 39
Metal Edge or Gravel Stop

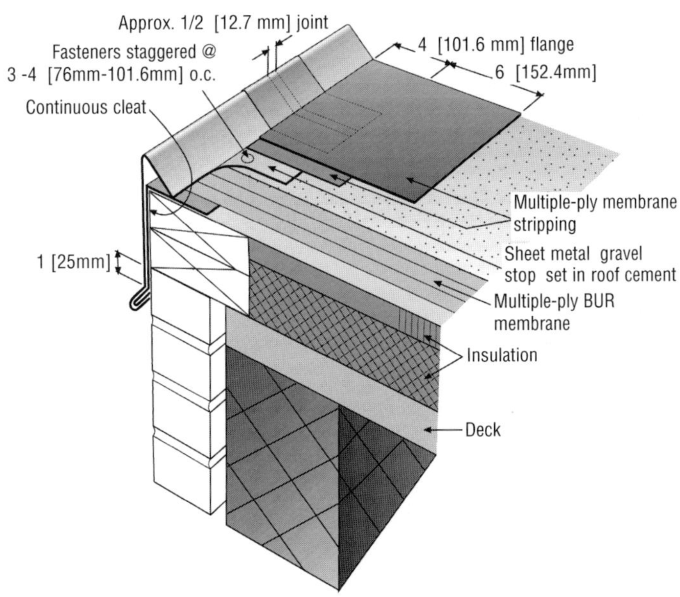

Light-gauge metal edges are subjected to high wind-uplift forces. In fact, most roof blow-offs begin at the building perimeter. Once the edge of the metal is lifted, the roof can peel back. To provide protection against blow-offs, an architect must specify a continuous cleat to restrain the vertical leg of the metal edge (figure 40). The installation and gauges of the metal edge and cleat should conform to FM s *Loss Prevention Data Sheet 1-49* (figure 41).

Extruded metal edges (figure 42) are too stiff to be anchored by closely spaced nails. They should be installed on raised edges above the membrane as discussed below and allowed to move due to thermal expansion and contraction of the metal. Since they are inherently stiff, they do not require cleats.

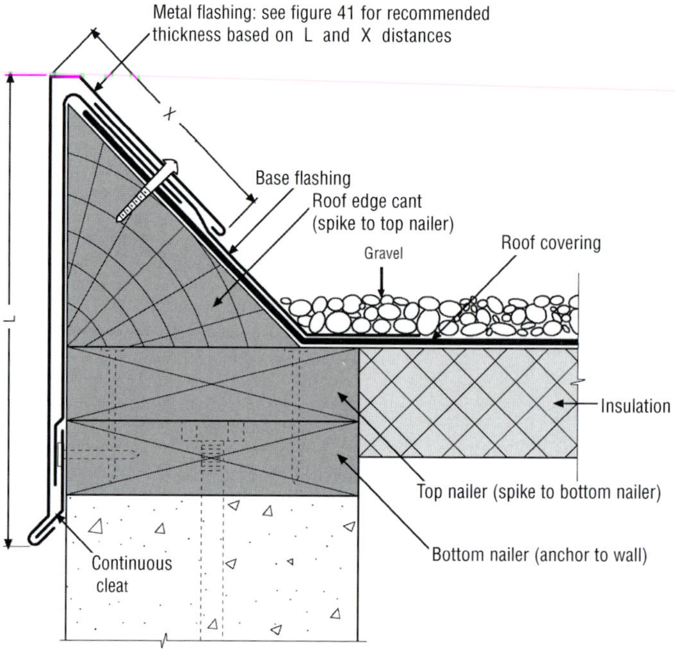

FIGURE 40

Metal Edge

SOURCE: Factory Mutual Research Corporation, FM Approval Guide, FM Loss Prevention Data Sheet 1-49

FIGURE 41

SOURCE: Factory Mutual Research Corporation, FM Approval Guide, FM Loss Prevention Data Sheet 1-49

Maximum Recommended "L" & "X" Dimensions and Metal Thicknesses for Metal Edge Flashing & Continuous Cleat*

TYPE OF METAL	MAXIMUM "L" DIMENSION			CONTINUOUS CLEAT	MAXIMUM "X" DIMENSION		
	VELOCITY PRESSURE (pounds per square foot) [kilopascal]				VELOCITY PRESSURE (pounds per square foot) [kilopascal]		
	WIND ZONE 1		WIND ZONE 2		WIND ZONE 1		**WIND ZONE 2
	10-20 psf [0.48-0.95 kPa]	21-30 psf [1.0-1.44 kPa]	31-45 psf [1.48-2.15 kPa]		10-20 psf [1.48-0.95 kPa]	21-30 psf [1.0-1.44 kPa]	31-45 psf [1.48-2.15 kPa]
Galv. Iron or Soft Stainless Steel Ga. [mm]	in. [mm]	in. [mm]	in. [mm]	Ga. [mm]	in. [mm]	in. [mm]	in. [mm]
26 [0.45]	6 [150]	6 [150]	4 [100]	24 [0.61]	3 [75]	2½ [65]	use thicker metal
24 [0.61]	8 [200]	8 [200]	6 [150]	22 [0.76]	4 [100]	3 [75]	3 [75]
22 [0.76]	10 [250]	10 [250]	8 [200]	20 [0.91]	4 [100]	4 [100]	3½ [88]
20 [0.91]	12 [300]	12 [300]	10 [250]	18 [1.06]	5 [130]	4½ [115]	4 [100]
Aluminum*** in. [mm]	in. [mm]	in. [mm]	in. [mm]	in. [mm]	in. [mm]	in. [mm]	in. [mm]
0.040 [1.02]	6 [150]	6 [150]	4 [100]	0.050 [1.27]	4 [100]	3 [75]	2½ [65]
0.050 [1.27]	8 [200]	8 [200]	6 [150]	0.060 [1.62]	4½ [115]	4 [100]	3 [75]
0.060 [1.62]	10 [250]	10 [250]	8 [200]	0.070 [1.78]	5 [130]	4½ [115]	3½ [88]
0.070 [1.78]			10 [250]	0.080 [1.93]	5½ [140]	5 [130]	4 [100]
Copper oz. [mm]	in. [mm]	in. [mm]	in. [mm]	oz. [mm]	in. [mm]	in. [mm]	in. [mm]
16 [0.55]	8 [200]	6 [150]	4 [100]	20 [0.69]	3½ [88]	3 [75]	2½ [65]
20 [0.69]	10 [250]	8 [200]	6 [150]	24 [0.82]	4 [100]	3½ [88]	3 [75]
24 [0.82]		10 [250]	8 [200]	32 [1.10]	4½ [115]	4 [100]	3½ [88]
32 [1.10]			10 [250]	48 [1.64]	5½ [130]	5 [130]	4 [100]

NOTES:

**1. Not applicable for metal gravel guard and fascia.*

***2. Additional fasteners are required near top of cant. One screw with washer at center of each section of fascia and one through joint cover.*

****3. Temper "O" aluminum, although easily formed, has a low bending strength. High tempers are advised when using aluminum.*

FIGURE 42
Extruded Metal at Raised Edge
SOURCE: NRCA

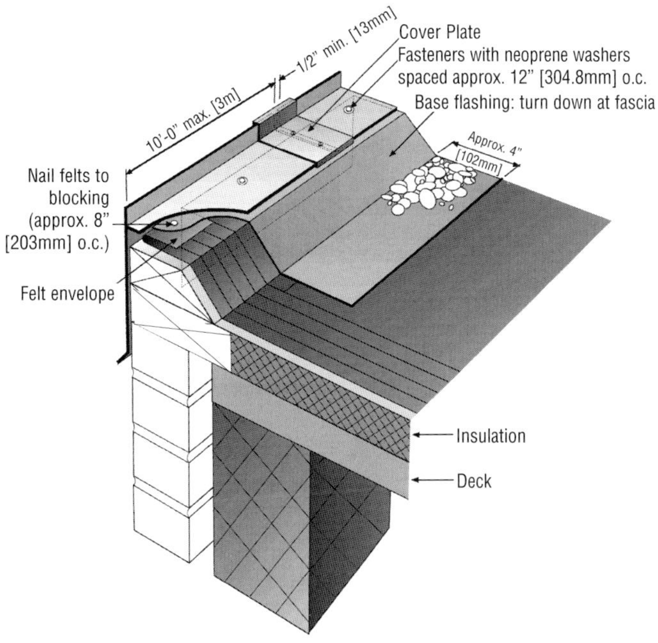

FIGURE 43
Raised Edge with Metal Flashing
SOURCE: NRCA

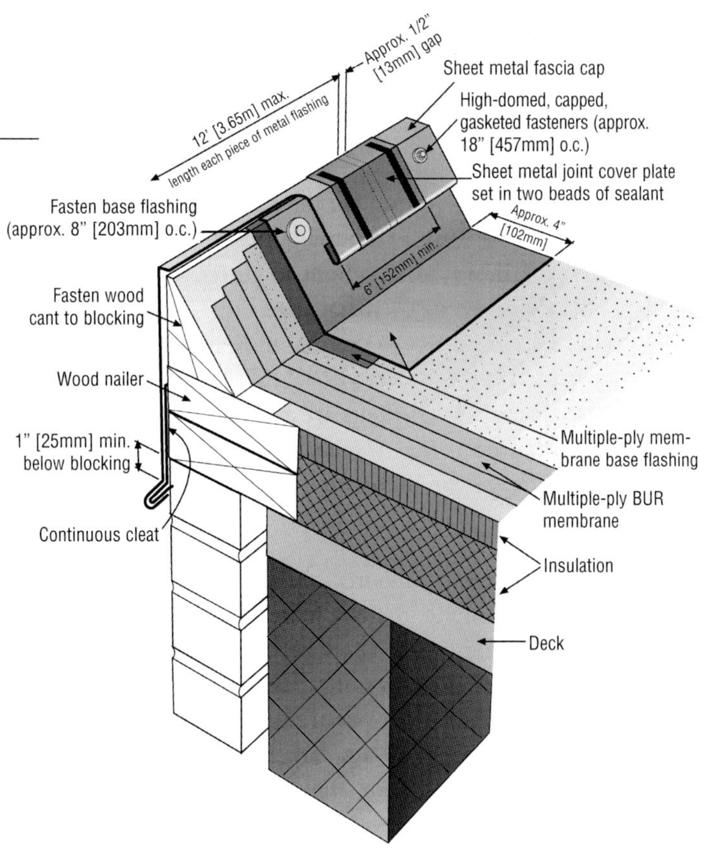

Application on a Raised Edge

A raised edge, or curb, is used when a low-profile fascia is desired (figures 42 and 43). It is built up of wood nailers, which are bolted securely to the underlying roof deck or wall. The membrane is then terminated at the top of this curb. A flashing felt, applied over the membrane, extends down over the face of the nailer and behind the metal edge. The light-gauge or extruded metal edge does not need stripping felts because it is above any water on the roof. The metal is free to move due to thermal changes and will not affect the membrane or flashing felts.

Raised edges are generally used with internal roof drains. If the roof slopes toward the perimeter, however, crickets are needed to direct the water to scuppers and downspouts. With the exception of small roofs, scuppers should be avoided in climates subject to freezing temperatures since slow-flowing water on large low-slope roofs can form icicles at these external drains. Scuppers used at roof edges should be shop-fabricated so that they are watertight. They should be open at the top to

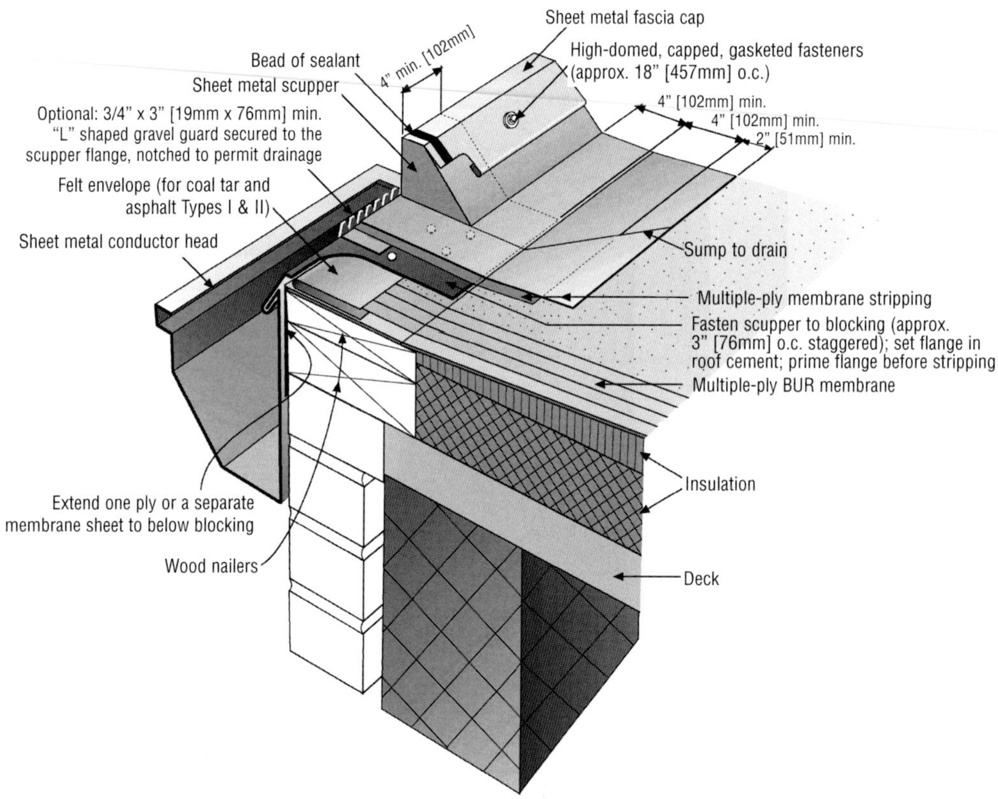

FIGURE 44
Scupper Through Roof Edge
SOURCE: NRCA

simplify flashing. The scupper flanges are stripped into the membrane with felts (figure 44).

Gutters

An interior drainage system is preferred to gutters on low-slope roofs. If they cannot be avoided, gutters should be hung from brackets on the outside of the building. Gutters should not be built into the eaves. A leak from the joint in a hung gutter is insignificant compared to the damage that can be caused by a leaking built-in gutter. Gutters should be formed or fabricated independently of the roof's metal edge.

Gutters are fabricated from materials that are subject to thermal movement. Expansion joints, therefore, should be provided every 40 feet [12.19 m]. Tables and guides for sizing gutters and for sizing and spacing downspouts are available from various sources, such as *The NRCA Roofing and Waterproofing Manual, Architectural Sheet Metal Manual* and *Ramsey/Sleeper Architectural Graphic Standards.*

PENETRATIONS

Penetrations, such as roof drains, skylights, equipment supports, pipes, scuttles and chimneys, disrupt the continuity of the roof membrane, thereby increasing the possibility of leaks. Penetrations must be located a minimum of 12 inches [304.80 mm] from the toe of cant strips and be spaced 12 inches [304.80 mm] clear of each other to provide sufficient clearance for the installation of flashing material (figure 45).

Penetrations can be accommodated in any number of ways, largely depending on their shape and size. Whenever possible, roof penetrations for supporting fencing, mechanical equipment, water tanks, signs and ductwork should be round in shape so that base flashings and counterflashings can be used. Avoid angles, wide-flange shapes,

FIGURE 45
Clearances
Provide adequate space between pipes and from walls and curbs.
SOURCE: NRCA

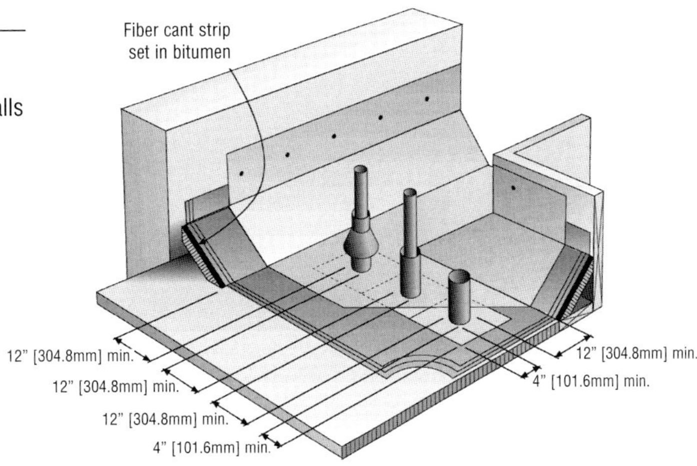

or square or rectangular tubes, which require the less desirable system of pitch pans. Large structural shapes can be mounted on concrete stub columns. Small angles that support ductwork can be mounted on precast concrete pavers that have been installed over the membrane. A curb with counterflashing is required for large penetrations.

FIGURE 46
Flashing for Plumbing Vent Stacks

Round Penetrations

When flashing round penetrations, consider the following general rules:

Plumbing vent stacks that measure 24 inches [609.60 mm] or less above the roof and are less than 6 inches [152.40 mm] in diameter should be flashed with 3-pound [1.36 kg] lead sheet formed as a sleeve. If the top of the stack is open, the metal should be turned down into the pipe or, preferably, covered with a second piece of metal that caps the pipe (figure 46).

Any vent stack that is higher or wider should be treated as a pipe.

Pipes that are between 4 and 8 inches [101.60 and 203.20 mm] in diameter should be flashed with metal base flashing that is covered by a metal hood (figure 47).

Pipes over 8 inches [203.20 mm] in diameter should be provided with a curbed opening in the roof and

FIGURE 47
Equipment Support Stand and Typical Rain Collar Penetration Detail
SOURCE: NRCA

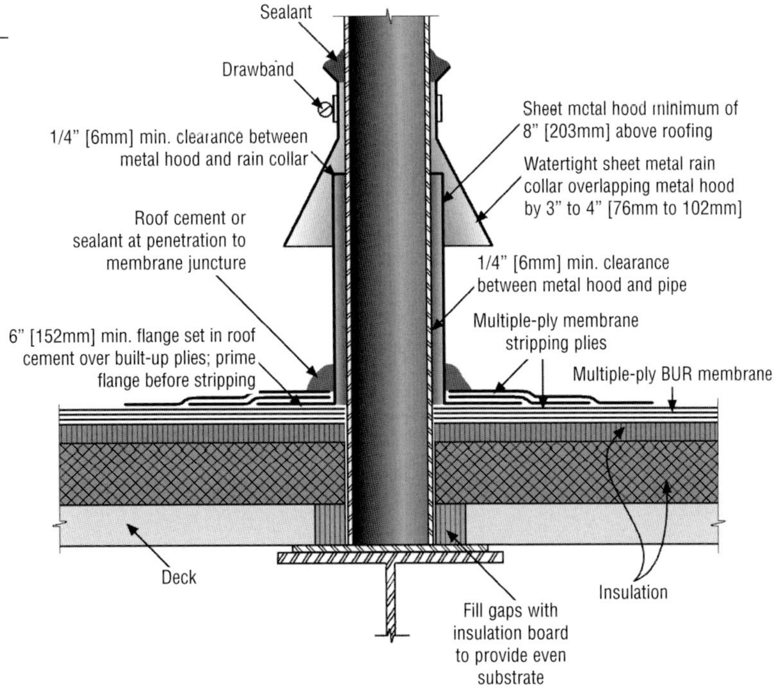

counterflashed with a metal hood (figure 48).

Metal base flashing on flues up to 8 inches [203.20 mm] in diameter and with surface temperatures above 180 F [82.00 C] should be held 2 inches [50.80 mm] away from the flue surface. For counterflashing, a metal hood with a diameter 2 inches [50.80 mm] greater than that of the base flashing should be used (figure 48).

Groups of conduits should be brought up through a curbed opening and turned 90 degrees through a metal hood (figure 49).

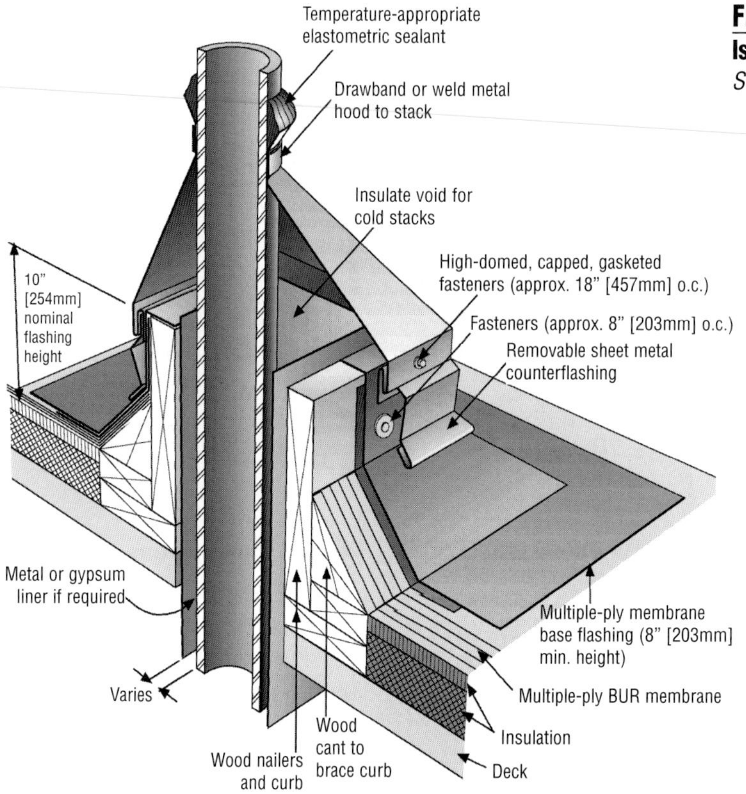

FIGURE 48
Isolated Stack Flashing
SOURCE: NRCA

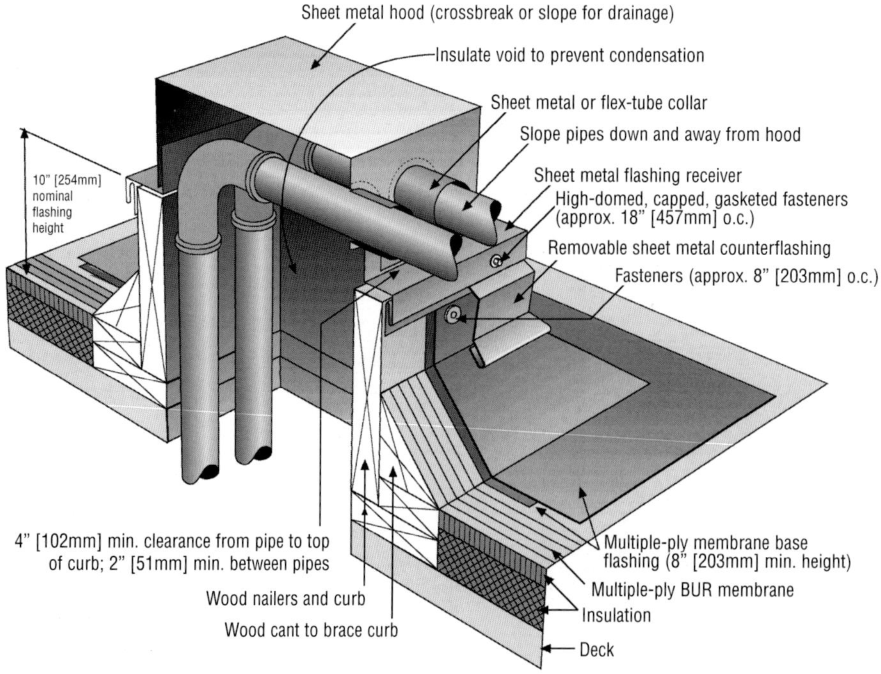

FIGURE 49
Curbed Opening for Groups of Pipes and Conduit
SOURCE: NRCA

FIGURE 50
Pitch Pans
SOURCE: NRCA

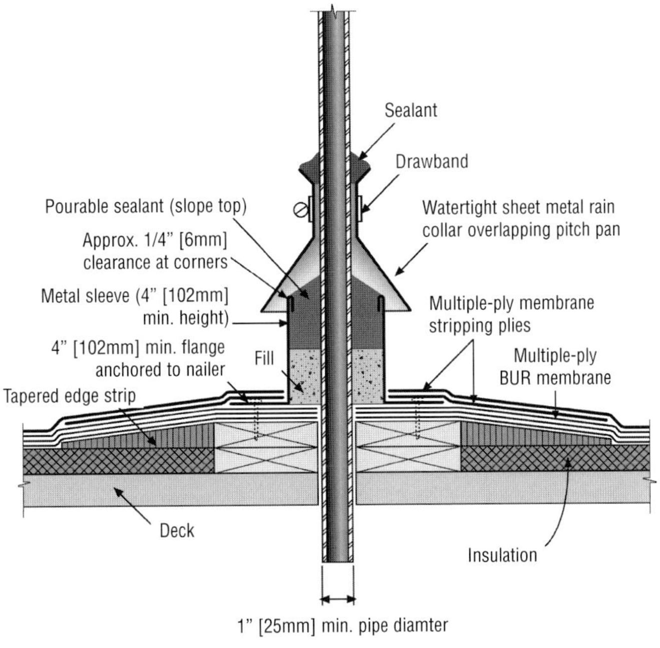

Pitch Pans

Pitch pans, or pitch pockets, are often used to flash irregular shapes that penetrate through a roof membrane. They have many disadvantages, including:

- They have frequent maintenance requirements.
- Separations between multiple penetrations must be larger than when curbed openings are used.
- The filler materials used in the pans are not long lasting.
- They are excluded from warranty programs.

When pitch pans cannot be avoided, they should be fabricated from noncorrosive metals, measure at least 4 inches [101.60 mm] high, have 4-inch-wide [101.60 mm] flanges and be secured with nails to treated wood-blocking located below the membrane (figure 50). The flanges of the pans are installed on the membrane with roofing cement and stripped in with two plies of flashing fabric or felts. Tapered edge strips should be used around the pan to provide drainage away from the pan. There should be a minimum of 2 inches [50.80 mm] between the edge of the pan and the item penetrating the pan. The pan is filled with polyurethane foam, roofing cement mixed with plaster, or cement to within 1 inch [25.40 mm] of the top. The remaining inch is filled with a two-part pourable sealer that is sloped from the center of the pan to the edges to provide positive flow of water away from the penetration. The pourable sealers are superior to bituminous fillers, which dry out and shrink with time.

FIGURE 51
Flashing for Large Penetrations
SOURCE: NRCA

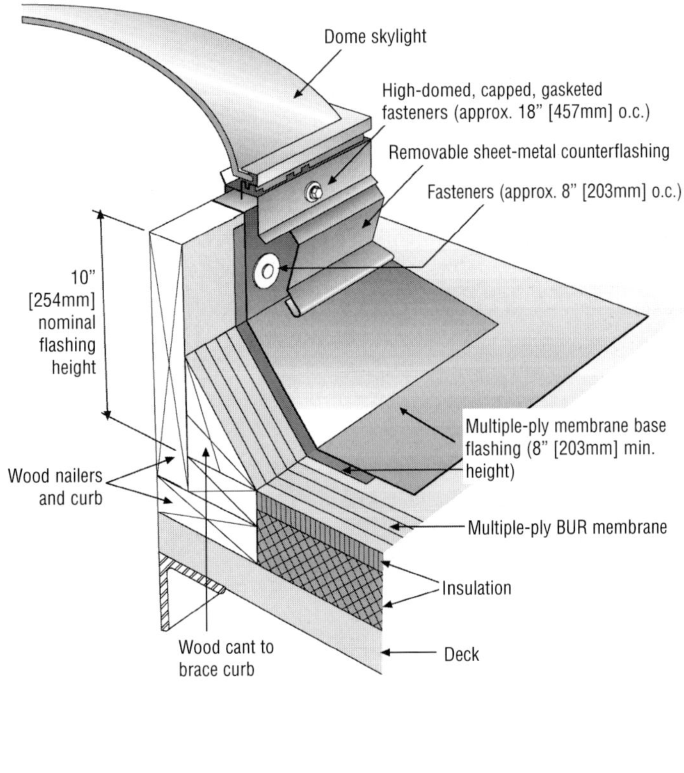

Curbed Penetrations

Large penetrations, such as ducts, fans, metal flues, scuttles and skylights, must be flashed with a curb and counterflashing (figure 51) similar to the system used on rising walls and parapets. The curb may be prefabricated or constructed on site with pressure-treated wood members.

The curb must form a box that rises at least 8 inches [203.20 mm] above the roof membrane. Cant strips must be provided. If the counterflashing does not cover the curb by at least 4 inches [101.60 mm], the base flashing plies should be carried over the top of the curb.

Small fans and skylights usually have adequate counterflashings. Chimneys and ductwork will require separate sheet-metal hoods. Some prefabricated curbs are provided with integral cants and mounting flanges. The flanges should be installed above the membrane, securely fastened to treated wood blocking and covered with base flashing as these curbs are rarely watertight.

Internal Roof Drains

Internal drains are roof penetrations that require particular care because so much water is flowing to them. Drains can be underwater at times so the flashings that connect them to the membrane must be capable of withstanding hydrostatic pressures. Drains are installed in a hole cut through the deck. In most cases, the drains are independent of the deck. This means that the connection between the drain and membrane must be able to absorb the vertical and horizontal forces caused by differential movement. Four-pound [1.81 kg] lead sheet is used to reinforce the flashing because it is malleable and can withstand these forces. A 30-inch-square [762.00 mm-square] lead sheet is installed in roofing cement over the membrane that has been brought into the drain. The lead is stripped into the membrane with roofing felts and then the lead and all the felts are clamped to the drain body by bolting the clamping ring to the drain (figures 52 and 53).

FIGURE 52
Lead Flashing at Drain Body

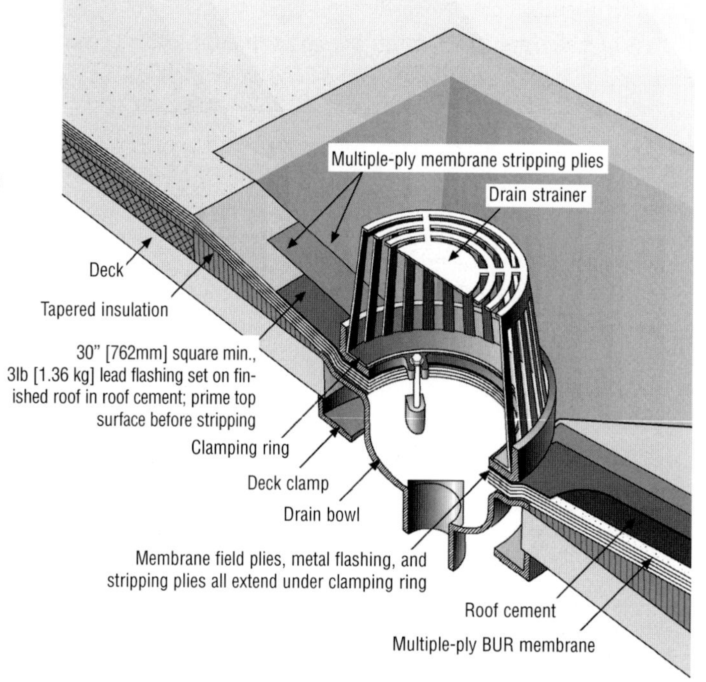

FIGURE 53
Roof Drain
SOURCE: NRCA, The NRCA Roofing and Waterproofing Manual—Fourth Edition

FIGURE 54
Flashing for Equipment Support Curb
SOURCE: NRCA

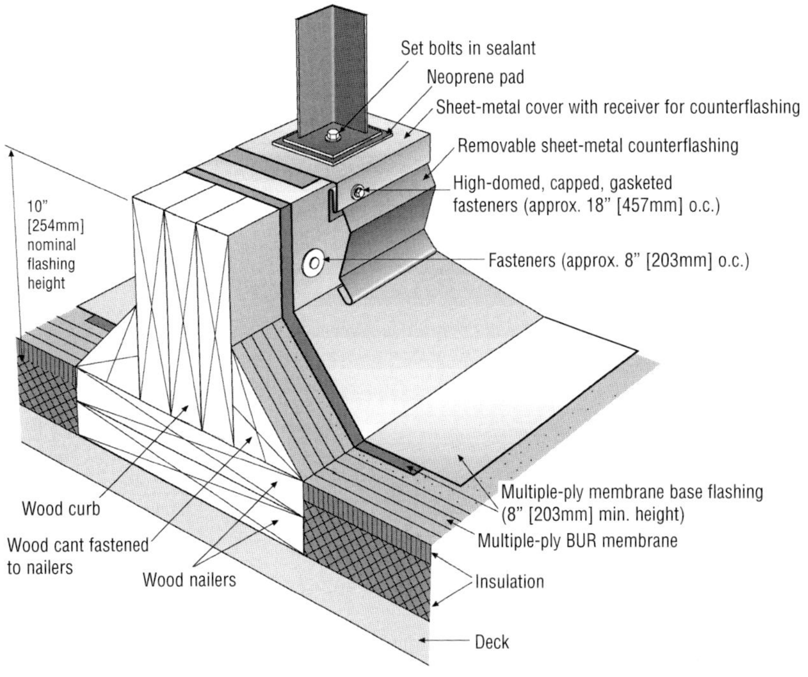

MECHANICAL EQUIPMENT

HVAC units, ductwork, air handlers, fans and gas piping are among the many mechanical items installed on roofs. Each of these items will require one or more roof penetrations, each of which must be flashed. Most of these items will also require regular maintenance that will generate roof traffic and sometimes abuse.

Too often roof leaks can be traced to a mechanic who punctured the membrane by dropping an access panel on it. To avoid these problems, mechanical equipment should be housed in penthouses. When this is not possible, roof equipment should be ganged and mounted on raised curbs (figure 54) or stands (figures 55 and 56). Several units grouped together can be easily screened.

When designing flashings for rooftop equipment, consider the fact that the mechanical equipment may be replaced several times during the life of the building and the roof may have to be replaced during the life of the mechanical equipment. Careful planning is required to accommodate the eventual replacement of roofing material or mechanical equipment without affecting the other. The mechanical equipment should be separated completely from the roofing membrane. Do not install HVAC units on wood sleepers that rest on the membrane. Mechanical equipment should be installed on steel equipment stands that are raised sufficiently high above the membrane so that the roofing can be removed and replaced

FIGURE 55
Equipment Support Stand
SOURCE: NRCA

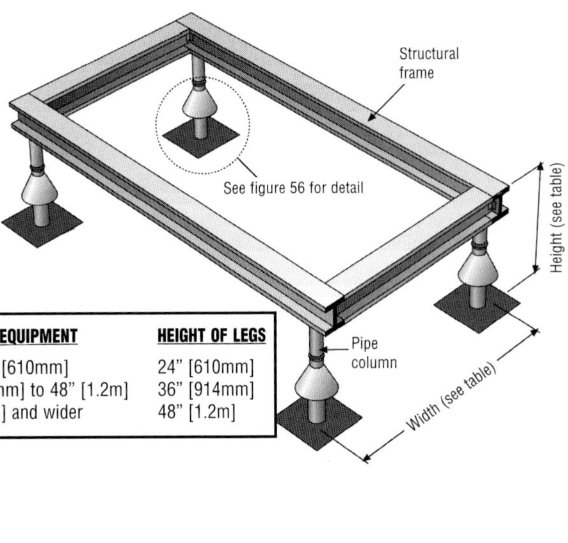

WIDTH OF EQUIPMENT	HEIGHT OF LEGS
Up to 24" [610mm]	24" [610mm]
25" [635mm] to 48" [1.2m]	36" [914mm]
48" [1.2m] and wider	48" [1.2m]

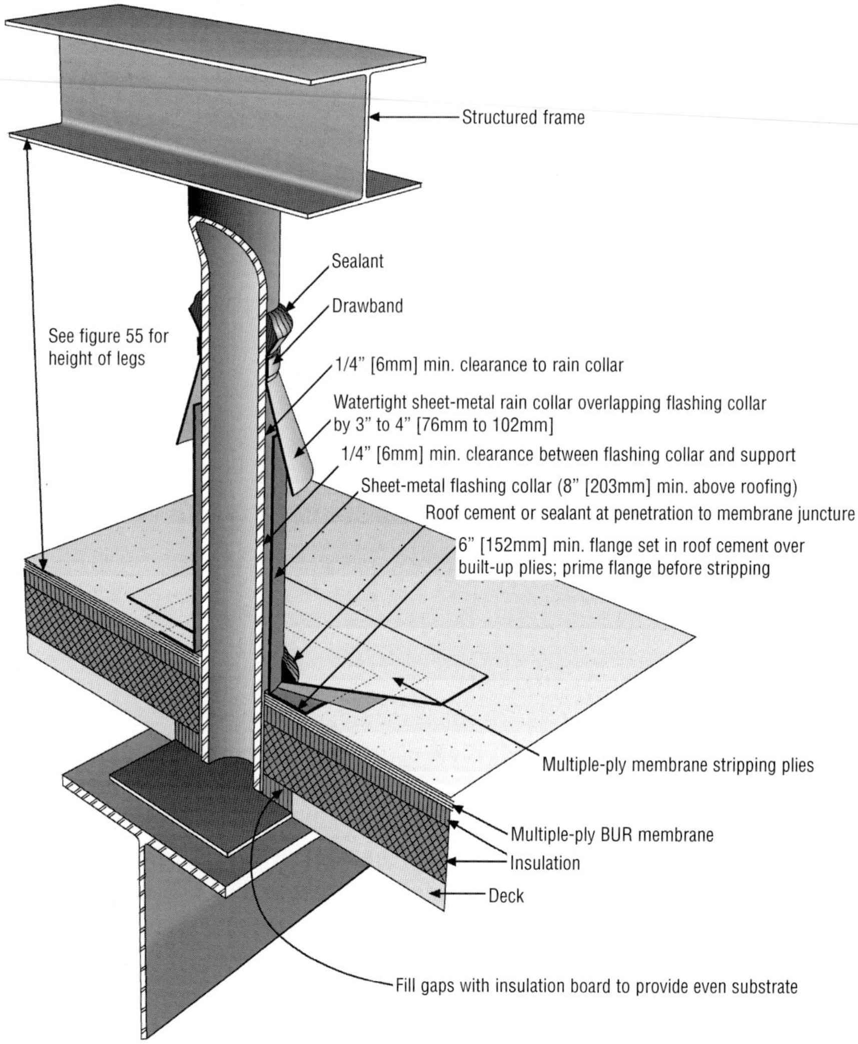

FIGURE 56
Flashing for Equipment Support Stand and Typical Rain Collar Penetration Detail
SOURCE: NRCA

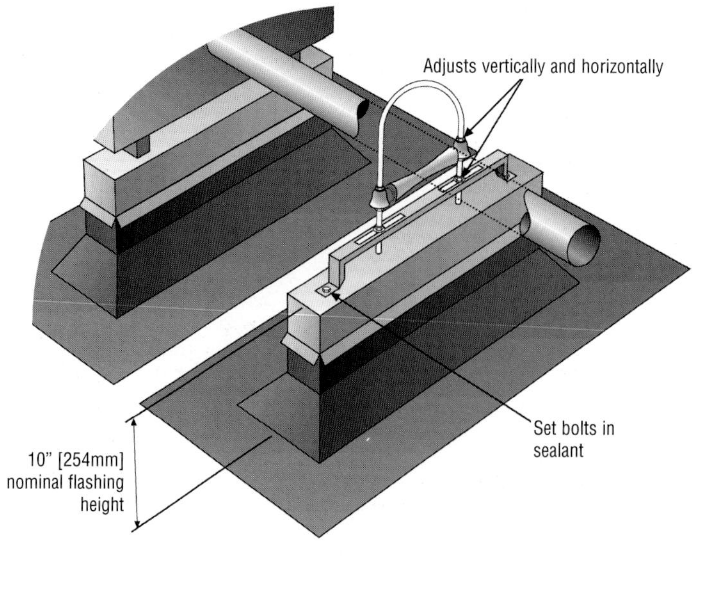

FIGURE 57
Pipe Roller Support
SOURCE: NRCA

FIGURE 58
Conditions Requiring Expansion Joints

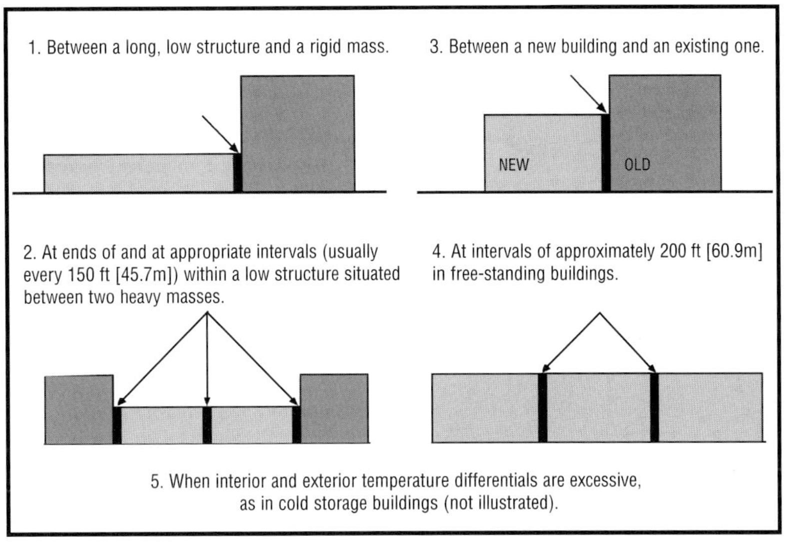

without moving the equipment. Equipment can also be installed over boxed curbs that provide access for flashing removal or replacement, if necessary. Do not install piping or conduit on wood blocks that rest on the membrane, but rather on raised pipe roller supports (figure 57). All piping and conduits should be flashed as discussed above for penetrations. Do not penetrate the base flashings on curbs. Condensate drains and overflows from HVAC units and cooling towers should be piped to roof drains and not allowed to flow over the roof surface.

EXPANSION JOINT COVERS AND AREA DIVIDERS

Movement or expansion joints are needed in buildings where cumulative or differential movement is expected (figure 58). These joints must be at least 1 inch [25.40 mm] wide, continuous through the building from foundation top to roof, and continuous through parapets and raised edges. When possible, locate the joints on high points of the roof so that water is draining away from the joints. Provide drains on both sides of joints that are carried through valleys.

Wood curbs at least 8 inches [203.20 mm] high are installed on either side of the joint and are flashed like parapet walls, described above. The top of the joint must be protected by a formed sheet-metal cover or an elastomeric material. The space between the curbs should be filled with flexible insulation to control heat loss and prevent condensation (figures 59 and 60).

Expansion joints

FIGURE 59
Expansion Joint with Metal Cover
SOURCE: NRCA

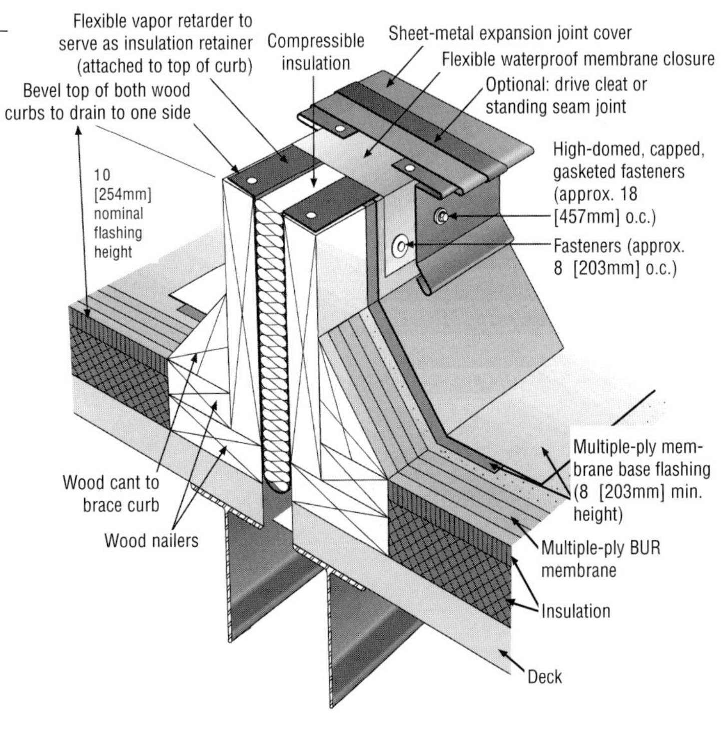

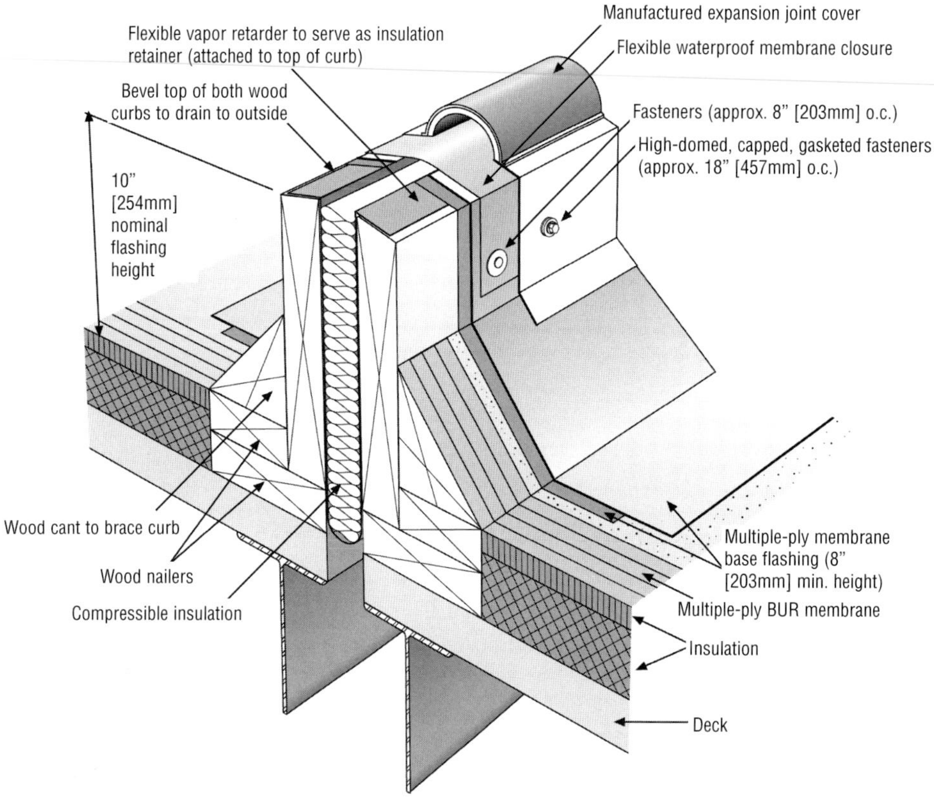

FIGURE 60
Expansion Joint with Manufactured Cover
SOURCE: NRCA

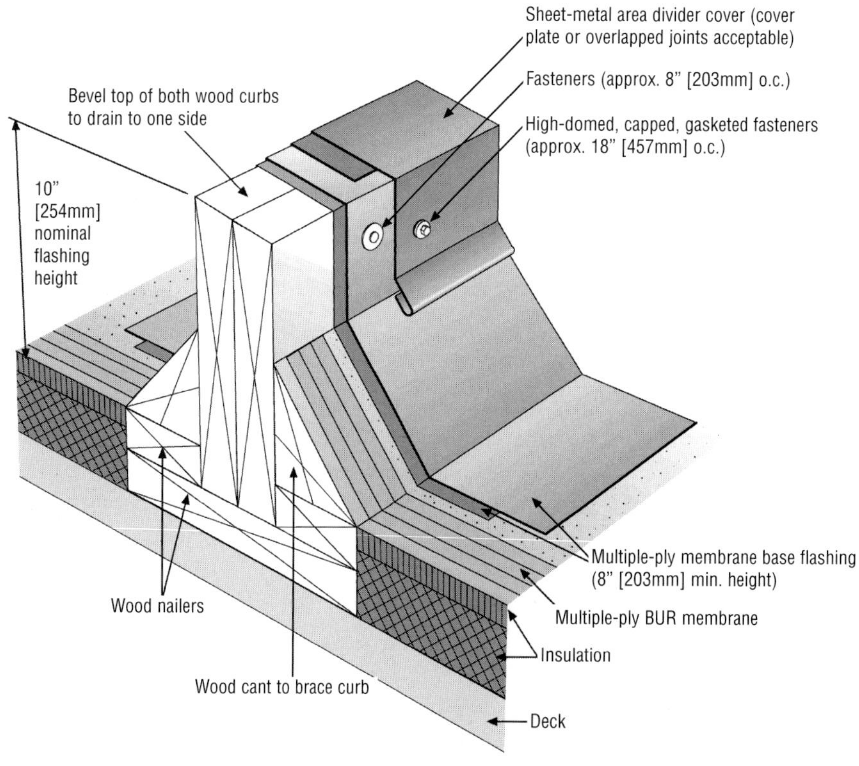

FIGURE 61
Area Divider
SOURCE: NRCA

are dynamic so failures of joint covers can be expected more often than that of base flashing. When joints are located above sensitive areas, such as laboratories, libraries and operating rooms, consider installing a gutter below the joint as a backup to the joint cover.

Area dividers (figure 61), which resemble expansion joint covers, are used to create smaller areas on large roofs. They are not located over expansion joints and are therefore not subject to movement. Area dividers are useful at reentrant corners on L-shaped buildings, at corners of penthouses and for accommodating a change in felt direction when installing the roof membrane.

IN THE FIELD

Having carefully researched the project's roofing requirements, selected the appropriate BUR system, and studied and drafted the roofing membrane termination details, the architect comes to the most difficult part of the process: observing the application. Most of the problems that occur in BUR systems are not the result of poor materials. Problems occur when materials are not selected correctly, when the use of materials is not clearly delineated in the drawings and specifications, and when the materials are not applied according to good roofing practice and judgment.

For the architect, construction observation is a form of quality assurance. Like any other part of the building construction, this entails meetings with all the involved parties to review contract requirements, procedures for auditing the contractor's quality control reports, field sampling to verify compliance and visual observations.

Some architects rely on the material manufacturer to provide inspection services. This is not recommended, however, since the inspector may also be the sales representative on the job and, therefore, have a conflict of interest. Also, these types of inspections may be infrequent, occurring only at the beginning and end of a job. If problems occur, the manufacturer will usually declare that their inspections were only to verify that their material was used and did not extend to the application procedure or quality. Moreover, the manufacturer's inspection is often limited to only those materials supplied by that company and not to other assembly components.

If properly trained to monitor roofing applications, an architect's field representative on the job site can fulfill this function. Since roofing application is the key to roofing performance, however, it may be in the best interest of the owner and architect to employ an outside consultant for inspection services in much the same way that outside services are used for concrete and steel applications. These services range from complete involvement, including roof design, to more limited observation and reporting during construction.

The inspector records the progress of the job, documents any areas that do not conform to the job requirements and guards the owner from defective workmanship. A checklist and a report form allow the inspector to record the observations in an orderly manner. The field inspector must be familiar with material protection, application requirements and protection of the roofing assembly during and after application.

MATERIAL AND STORAGE

Bituminous products should be delivered in containers that are clearly labeled with the name of the manufacturer, ASTM designation, type of bitumen, flash point and EVT. If required, UL and FM approvals should also be included. If the asphalt or coal-tar pitch is delivered in tankers, the shipment should be accompanied by a certificate that contains the same type of information. The architect should request similar labels or certificates for other materials, such as insulation, surfacings, accessories, mastics and fasteners.

All materials must be stored on pallets above ground and covered with tarpaulins, not plastic wrappers that can trap moisture (figure 62). Plastic wrappers that cover insulation and rolls of felt during shipping must be removed promptly. When insulation products absorb moisture, they lose their thermal resistance and contribute to roofing problems like blistering and ridging. In cold weather, felts and other temperature-sensitive materials should be stored in closed, heated storage trailers. Rolls of felt must be stored on end (figure 63). They will develop flat spots, which can result in application problems, if stored on their sides (figure 64).

Material that becomes wet or has not been properly stored must be marked and removed from the job. This should be done as soon as possible so that the contractor can order replacement material.

FIGURE 62
Insulation
Remove plastic shrink wrapping before storing insulation.

FIGURE 63
Felt Rolls
Felt rolls should be stored on their ends on top of pallets.

FIGURE 64
Felt Rolls
Do not store rolls on their sides.

DECK PREPARATION

The deck must be ready before roofing can begin. This includes installing all wood blocking, curbs for roof openings, supports for equipment stands, roof drains, expansion joints and scuppers. Without these accessories in place, the applicator cannot do a complete job and later problems are almost certain to occur.

The deck itself must be clean, dry and free of debris that could damage roofing felts. Concrete decks must be free of fins and ridges that would impede insulation installation. Chemicals that could react with bituminous adhesives must also be removed.

Precast decks must be checked for alignment (figure 65). Joints may have to be grouted or a 2-inch [50.80 mm] concrete topping applied to provide a level surface for the roofing application.

Steel decks must be checked for proper attachment. End laps must be nested and side laps fastened. Welds must be checked at supports and mechanical fasteners installed where needed.

Wood and plywood decks need a layer of rosin-sized paper to separate the deck from the bituminous materials in the roof membrane.

FIGURE 65
Precast Deck
Check alignment to ensure level surface.

Check fasteners to ensure proper attachment to the structural members below the deck. Also, check that large openings in the roof deck have proper structural supports and that provisions have been made for the support of rooftop equipment.

INSTALLING THE INSULATION

Roofing application is a continuous operation that starts with insulation and finishes with the surfacing. No more insulation may be installed than can be completely covered by the finished membrane and flashings on the same day.

The weather must be suitable for any roofing to begin. Weather conditions must be reported on the inspection forms. Roof problems can develop later if installation takes place during improper weather conditions.

FIGURE 66
Attaching Insulation to Steel Deck with Mechanical Fasteners

If mechanical fasteners are to be used (figure 66), a shop drawing must be on the job that shows the number and location of fasteners in each insulation board. Additional fasteners are needed in each board around the perimeter and at the corners of the building. Check the fasteners against the shop drawings for correct size, spacing and length.

FIGURE 67
Unsupported Insulation Edge
This is not acceptable.

Insulation installed on a steel deck must be fully supported along the edges parallel to the deck flutes. It cannot be allowed to cantilever over the flutes (figure 67). If the insulation dimensions do not agree with the flute spacing, the insulation will have to be trimmed so that full bearing is achieved or metal strips must be installed to cover the flute and support the insulation.

Concrete decks need to be primed before insulation is attached in hot bitumen. The deck must also be dry for proper adhesion to occur. The deck is too wet if hot bitumen bubbles as it is applied or if the bitumen can be easily peeled off.

MEMBRANE APPLICATION

Roofing membranes are fabricated in the field by alternatively applying layers of hot bitumen and reinforcing felts. The materials may be applied by hand using a mop or bitumen dispenser and manually rolling the felt into the hot bitumen. Alternatively, a felt-laying machine can dispense the hot bitumen and unroll the felt into it (figure 68).

FIGURE 68
Felt-laying Machine

The goal of either method is to completely and fully adhere each layer of felt in a complete film of hot adhesive. Felt should not touch each other.

FIGURE 69
Squeegee for Application of Glass Felt

Each felt should be fully adhered to the felt below with hot adhesive. Adhesion is achieved by using a hand-held push broom to smooth out and embed the felt into the hot material. When glass felts are installed, a squeegee (figure 69) is used instead of a broom.

Hand applications are useful on small areas and on roofs with many penetrations. Machine applications are useful on a large, open roof with long runs. Union rules may prohibit these machines in some areas. If the machine is properly calibrated, its moppings are more uniform than those achieved by hand. Machines must be periodically checked to ensure that a uniform film of hot bitumen is being applied between the layers of felt. Long voids, which are the beginnings of blisters in the system, can result if the openings in the dispenser are clogged.

The layers of felt in a BUR application are installed in shingle style (figures 70 and 71). Laying lines are preprinted on the face of the felt by the manufacturer to assist the applicator in providing the proper headlaps and the correct number of felts at any place on the roof. Unfortunately, the laying lines are not always accurate, so

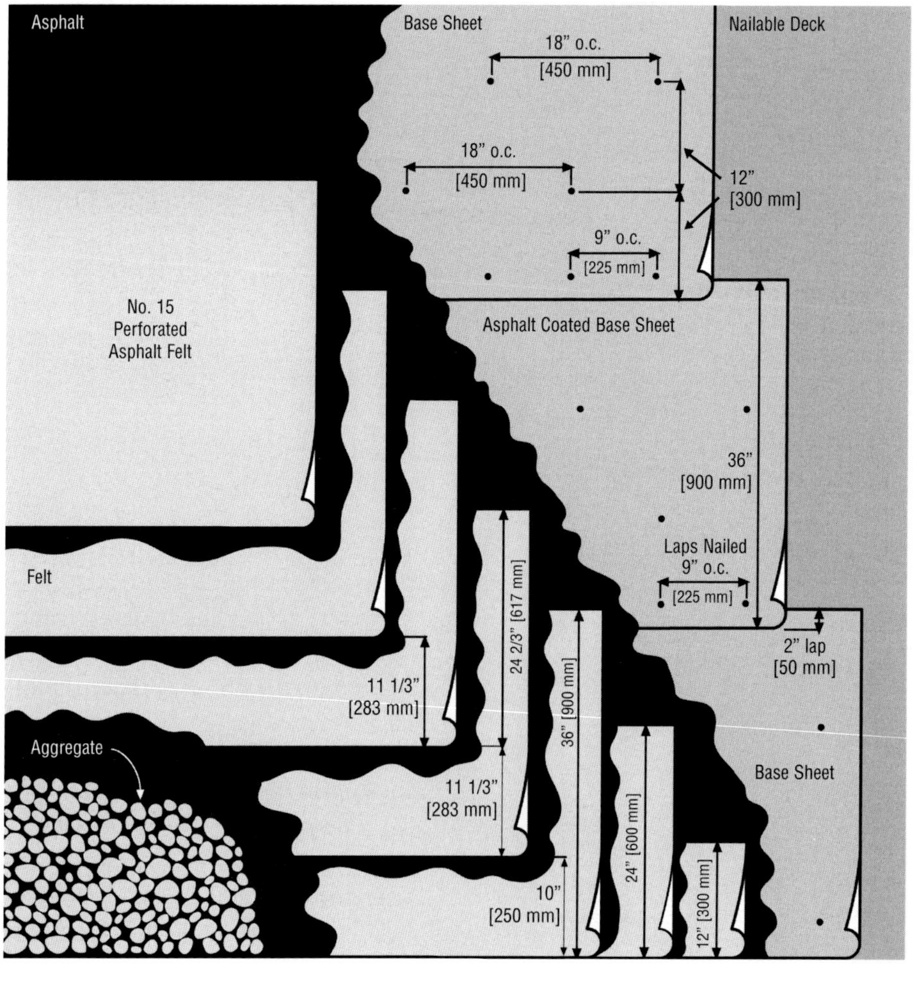

FIGURE 70
Shingle-style Application of Roofing Felts
An aggregate surfacing system is shown.

FIGURE 71
Shingle-style Application of Roofing Felts
A mineral-surfaced cap sheet system is shown.

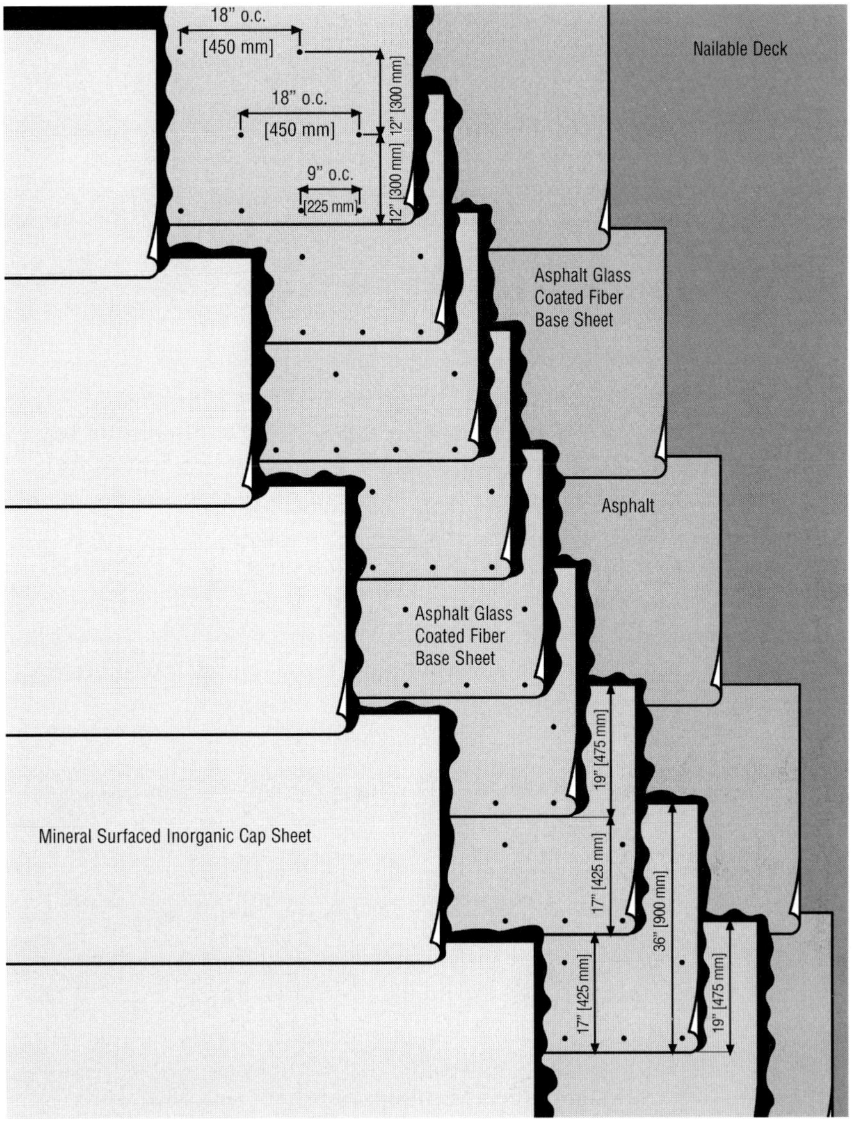

FIGURE 72
Bitumen Runout Between Plies

felt exposure must be checked periodically. It may be necessary to mark each layer of felt with a chalk line to ensure proper lapping of the material.

The hot bitumen between each layer of felt must be uniform and at the proper temperature. Otherwise full adhesion will not occur and blisters may develop. A small amount of mopping material should be clearly visible, running beyond the edge of the felt (figure 72). All of the roof plies must be applied at the same time: The practice of phasing, in which some felts are applied on one day and others at a later date, is not acceptable. Phasing creates a slippage plane between the felts. It also can result in dirt accumulation on the exposed felts, which will inhibit adhesion to the subsequent ply. Any defects, such as fish mouths (figure 73), blisters, or splits, must be repaired immediately.

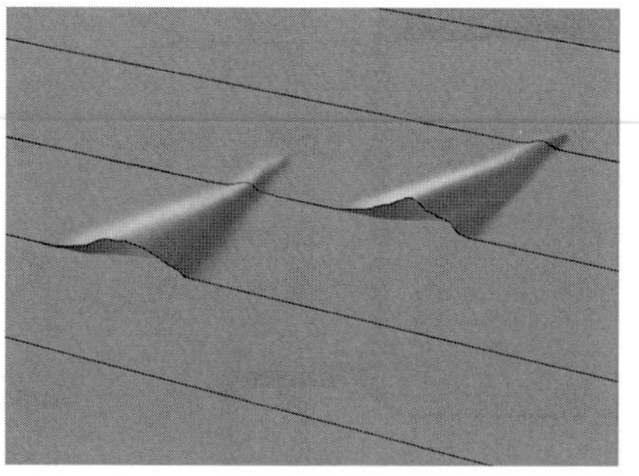

FIGURE 73
Fishmouths

At the end of a day's work the roof system must be sealed to the deck so that water cannot get under the membrane or into the insulation. This is a relatively easy task on concrete decks. It is also relatively easy on steel decks when the long edges of the insulation run parallel to the deck's flutes. On steel decks where the long edges of the insulation are perpendicular to the flutes, however, the flutes must be filled with pieces of insulating material to form a watertight seal.

A note of caution: Never allow the applicator to walk on or roll mop carts across glass felts as they are being installed. No traffic should be allowed until the asphalt has cooled and hardened. Glass felts are porous. Any traffic over them during installation can squeeze the hot bitumen through the mat, resulting in voids between the felts. In such cases, water could permeate completely through the membrane and show up as a leak in the space below.

Ideally, no more roofing should be installed than can be completed—with flashings and surfacings—in one day. When it is not possible to apply the surfacing, the felts must be glazed with a thin coat of asphalt to protect them from water. Nonetheless, this practice leaves the roof surface open to traffic damage. It should be carefully inspected and repaired if necessary before the aggregate and flood coating are applied.

FLASHINGS AND ACCESSORIES

Flashings attach metal, masonry or other surfaces to bituminous products. More problems and failures occur at flashings than at any other part of the BUR system.

A primer is necessary to prepare the surface to accept the bituminous materials. Make sure that all of the required nailers are in place and that they are made of pressure-treated wood.

Since flashings generally provide the transition from a horizontal to a vertical surface, they are subject to the force of gravity. Their tops, therefore, must be secured to prevent sagging and failure. They must also be sealed at the top from water. Counterflashing is designed to protect the top of the base flashing but not to make it watertight. Counterflashings are made of metals that expand and contract continuously. They will buckle and fail if provision is not made in their design and installation for them to move.

The most difficult areas to flash and counterflash are where flashings terminate against vertical surfaces such as parapets or meet other flashing systems. Check shop drawings very carefully to ensure that proper transitions have been provided and that it is not left up to the installer to innovate in the field. Field innovations usually end up as future problems.

SURFACING

Aggregate surfacing is installed in hot flood coats of bitumen after the membrane and flashings have been installed. The aggregate must be reasonably dry or it will not adhere to the bitumen. If a mineral-surfaced cap sheet is used as the final surfacing, the rolls of materials must be unrolled and left to relax before they are installed. If the weather is cold, the rolls must be stored in warm storage buildings or trailers so that they will unroll without cracking and becoming damaged. If a reflective coating is to be applied, the applicator must wait until the roof surface has been exposed for several weeks. Failure to wait will cause the roofing asphalts to bleed through the reflective coating, making them useless as a reflective surface or as a decorative element.

PROTECTION

Once the final surfacing is applied, every effort should be made to keep traffic off the roof. Many problems in roof assemblies can be traced to abuse by other trades on the roof after it has been applied. Masons and ironworkers working above lower roof surfaces are prone to dropping materials that can seriously damage the roof assembly. Transporting materials and equipment over a finished roof is a sure way to have later leaks and problems.

If the roof must be used as a working platform, then protection must be provided. This can be in the form of plywood runways and barriers to keep traffic confined to certain areas. Temporary ramps must be provided at expansion joints so that equipment transported on wheeled carts can cross the joints without damaging them. Care should be taken not to overload the deck with equipment, storage of materials, or debris. This is particularly important on steel decks where excess loading can result in permanent deformation of the deck and its supports.

FINAL INSPECTION

It is a good idea to delay the final inspection and acceptance of any roof assembly until all trades are off the roof and the roofing applicator states that all of his work is complete. The roofing contractor normally wants to have his final inspection immediately after his work is complete so that any damage done by other trades can be claimed as extra work. This is something that the general and roofing contractor can work out between themselves. The architect's job is to ensure that the owner gets a complete, watertight job; this won't be known until all of the work by all the trades is completed.

Inspect the roof as if it has been in service for a year or more. Check every flashing, seam and accessory for movement or rupture of flashing felts. Most of the damage from other trades will be at the flashings. Where leaks have occurred or major problems are suspected, a nondestructive moisture test, such as infrared thermography, nuclear moisture meters or capacitance devices, may be appropriate to determine if any water has penetrated the system.

ABS	acrylonitrile butadiene styrene
ACI	American Concrete Institute
AIA	The American Institute of Architects
APA	APA—The Engineered Wood Association
ARMA	Asphalt Roofing Manufacturers Association
ASCE	American Society of Civil Engineers
ASHRAE	American Society of Heating, Refrigerating and Air-Conditioning Engineers
ASTM	American Society for Testing and Materials
BUR	built-up roofing
CIB	International Council for Building Research Studies and Documentation
CSI	Construction Specifications Institute
EPDM	a synthetic elastomer based on ethylene, propylene and diene.
EVT	equiviscous temperature
FM	Factory Mutual Research Corporation
NCARB	National Council of Architectural Registration Boards
NRCA	National Roofing Contractors Association
PVC	a synthetic thermoplastic polymer prepared from vinyl chloride
RIEI	Roofing Industry Educational Institute
SDI	Steel Deck Institute
SMACNA	Sheet Metal and Air Conditioning Contractors' National Association
UL	Underwriters Laboratories

Aggregate
Granular material such as gravel, crushed stone or slag used for surfacing.

Alligatoring
Shrinkage cracking of the bituminous surface of built-up roofing, or the exposed surface of smooth-surfaced roofing, in which the loss of volatile oils and the oxidation of compounds under solar radiation produces a pattern of deep cracks with the scaly look of an alligator's hide. It occurs only in unsurfaced bitumen exposed to the weather.

Asphalt
A dark brown to black, highly viscous, hydrocarbon produced from the residuum left after the distillation of petroleum. Used as the waterproofing agent of a built-up roof. It comes in a wide range of viscosities and softening points, from about 136 F [57.70 C] for dead-level asphalt to 210 F [98.80 C] or higher for special steep asphalt (see bitumen).

Asphalt Felt
An asphalt-saturated felt.

Backnailing
Blind nailing of roofing felts to provide additional slippage resistance.

Base Sheet
A saturated, impregnated or coated felt placed as the first ply in a multi-ply built-up roofing membrane.

Base Flashing
(1) The upturned edges of a membrane where it translates to a vertical plane at terminations and penetrations. (2) The metal or composition flashing used at the joint between the roofing surface and a vertical surface.

Bitumen
(1) A class of amorphous, black or dark-colored, (solid, semisolid, or viscous) cementitious substances, natural or manufactured, composed principally of high molecular weight hydrocarbons, soluble in carbon disulfide, and found in asphalts, tars, pitches, and asphaltities. (2) A generic term used to denote any material composed principally of bitumen. (3) In the roofing industry, there are two basic bitumens: asphalt and coal-tar pitch. Before application they are either (a) heated to a liquid state, (b) dissolved in a solvent or (c) emulsified.

Bituminous Emulsion
A suspension of minute particles of bituminous material in water or in an aqueous solution.

Blister
A spongy, raised portion of a roofing membrane, ranging in size from 1 inch in diameter and barely detectable in height to as much as 50 square feet [4.60 m2] in area and 1 foot [.30 m] high. Blisters result from the pressure of entrapped air or water vapor. Longitudinal blisters are called tunnel blisters and those that are non-linear are called mole runs.

Bond
The adhesive or cohesive forces preventing delamination of two roofing components (see also manufacturer's bond).

Built-up Roofing (BUR) Membrane
A continuous, semi-flexible multi-ply roof membrane consisting of plies or layers of saturated felts, coated felts, fabrics, or mats between which alternate layers of bitumen are applied. Built-up roof membranes are surfaced with mineral aggregate and bitumen, a liquid applied coating or a granule-surfaced cap sheet.

Cant Strip
A continuous beveled or triangular strip of wood, wood fiber, perlite or other material designed to serve as a gradual transition between the horizontal surface of the roof deck or insulation and a vertical surface.

Cap Flashing
(see counterflashing)

Cap Sheet
A granule-surfaced, coated sheet used as the top ply in a built-up roofing membrane.

Coal-tar Pitch
A dark brown to black, solid cementitious material obtained as residue in the partial evaporation or distillation of coal tar.

Coal-tar Felt
A felt saturated with refined coal tar.

Coated Base Sheet (or Felt)
A felt that has previously been saturated (impregnated with asphalt) and later coated with harder, more viscous asphalt, which greatly increases its impermeability to moisture.

Cold Flow
Slow deformation, under gravitational force, at or below room temperature.

Cold-process Roofing
A bituminous membrane comprising layers of coated or saturated felts bonded with coal-applied asphalt roof cement and surfaced with a cutback or emulsified asphalt roof coating.

Counterflashing
Sheet metal member that covers the top of the base flashing.

Creep
The dimensional change with time of a material under load, following the initial instantaneous elastic deformation.

Cricket
A raised diamond shaped segment of roof deck positioned between drains to provide a positive flow of water.

Cutback
Solvent-thinned bitumen used in cold-process roofing adhesives, flashing cements, and roof coatings.

Cutoff
A detail designed to prevent lateral water infiltration into the insulation where it terminates at the end of the day's work. Typically, a felt strip is hot-mopped to the stepped contour of the deck, the insulation edge, and the horizontal insulation surface.

Dead Level
Absolutely horizontal, of zero slope (see slope).

Deck
The structural substrate of a roof to which the roofing or waterproofing system (including insulation) is applied.

Delamination
A built-up roofing membrane failure characterized by separation of the felt plies, sometimes resulting in wrinkling and cracking.

Dry (n.)
A material that contains no more water than one would find at its equilibrium moisture content (see equilibrium moisture content).

Edge Stripping
Membrane flashing strips, cut to specific widths, used to seal perimeter edge metal to the roof membrane.

Edge Venting
The practice of providing regularly spaced openings at a roof perimeter to relieve the pressure of water vapor entrapped in the insulation.

Emulsion
A dispersion of fine particles or globules of liquid in a liquid. Asphalt emulsions consist of asphalt globules, an emulsifying agent such as bentonite clay and water.

Envelope (or Bitumen Stop)
A continuous edge seal formed at the perimeter and at penetrations by extending the base sheet or one ply of felt beyond the edge of the membrane plies. The extended ply is turned back onto the membrane and adhered. The envelope prevents bitumen seepage from the edge of the membrane into the building interior.

Equilibrium Moisture Content
Moisture content, expressed as percentage of moisture weight to material weight at a given temperature and relative humidity.

Equiviscous Temperature
Temperature at which the bitumen is at the most desirable viscosity for hot mopping and adhering membrane components.

Expansion Joint
A structural design detail that accommodates independent movement of building components.

Exposure
The transverse dimension of a felt not overlapped by an adjacent felt in a built-up roofing membrane. The exposure is thus that part of the felt covered directly by the flood coat. The correct felt exposure in a shingled, multi-ply roof is computed by dividing the felt width minus 2 inches [50.80 mm] by the number of plies. For example, for two plies of 36-inch-width [914.40 mm] felt, the exposure is (36 inches - 2 inches)/2 = 17 inches [(914.40 mm - 50.80 mm)/2 = 421.80 mm].

Fabric
A woven cloth; organic or inorganic filaments; threads or yarns.

Fallback
A reduction of bitumen softening point, sometimes caused by mixing asphalt with coal-tar pitch or overheating the bitumen (see softening point drift).

Felt
A flexible sheet manufactured by the interlocking of fibers through a combination of mechanical work, moisture, and heat, without spinning, weaving, or knitting. Roofing felts are manufactured from vegetable fibers (organic felts), glass fibers (glass-fiber felts), or polyester fibers (synthetic-fiber mats).

Flash Point
The temperature at which vapors from asphalt will ignite.

Flashing
The system used to seal membrane edges at walls, expansion joints, drains, gravel stops, and other places where the membrane is terminated or penetrated. Base flashing covers the edges of the membrane. Cap flashing, or counterflashing, shields the upper edges of the base flashing.

Flood Coat
The top layer of bitumen poured on a membrane into which aggregate is embedded.

Glass Felt
Glass fibers bonded into a sheet with resin and suitable for impregnation in the manufacture of bituminous waterproofing, roofing membranes, and shingles.

Glaze Coat
A thin coating of bitumen applied to the membrane as the top surfacing or when required for a temporary protective coat.

Heat Release
The amount of heat generated when an assembly is subjected to fire.

Live Loads
All loads, including snow, exerted on a roof except dead, wind, and lateral loads

Membrane
A flexible or semiflexible roof covering or waterproofing whose primary function is the exclusion of water.

Metal Flashing
A type of flashing frequently used as through-wall cap flashing, or counterflashing (see flashing).

Mopping
Application of hot bitumen with a mop or mechanical applicator to the substrate or to the plies of a built-up roof membrane. There are four types of mopping: (1) solid, in which the coating is continuous; (2) spot, in which the bitumen is applied in roughly circular areas, generally about 18 inches [457.20 mm] in diameter; (3) strip, in which bitumen is applied in parallel bands, generally 8 inches [203.20 mm] wide and 12 inches [304.80 mm] apart; and (4) sprinkle, in which bitumen is shaken on the substrate from a broom or mop in a random pattern.

Perm (Vapor Transmission)
A unit to measure water-vapor transmission. Its units are one grain of water vapor per square foot per hour per inch of mercury pressure difference.

Permeability
(1) The capacity of a porous medium to conduct or transmit fluids. (2) The amount of liquid moving through a barrier in a unit time, unit area, and unit pressure gradient not normalized for but directly related to thickness.

Permeance
The rate of water vapor transmission per unit area at a steady state through a membrane or assembly.

Phased Application
The installation of a roofing system during two or more separate time intervals.

Ply
A layer of felt in a roofing membrane; a four-ply membrane should have at least four plies of felt at any vertical cross section cut through the membrane.

Ponding
Water in low or irregular areas of a roof that remains longer than 48 hours after the cessation of rainfall.

Prime Coat
The first liquid coat applied in a multiple-coat system.

Prime (Bituminous)
A thin liquid bitumen applied to a surface to improve the adhesion of heavier applications of bitumen and to absorb dust.

Ridging
A failure of the membrane characterized by long blisters. When it occurs at insulation joints it is referred to as picture framing.

Roof Assembly
An assembly of interacting roof components, including the roof deck, vapor retarder (if present), insulation, and roof covering or membrane.

Roof Curb
An accessory used to mount and level equipment, such as exhaust fans or air conditioning units.

Roof System
A roof assembly excluding the deck.

Scupper
A channel through a parapet, designed for peripheral drainage of the roof. Typically used as a safety overflow to limit accumulation of ponded rainwater caused by clogged drains.

Slope
The angle or inclination of the roof from the horizontal expressed in inches of rise per lineal foot or occasionally in degrees.

Smooth-Surfaced Roof
A roof membrane without mineral aggregate surfacing.

Solid Mopping
See mopping

Spot Mopping
See mopping

Sprinkle Mopping
See mopping

Strip Mopping
See mopping

Through-wall Flashing
A water-resistant membrane or material assembly extending totally through a wall and its cavities, positioned to direct any water within the wall to the exterior.

Vapor Pressure
The pressure exerted by a vapor that is in equilibrium with its solid or liquid form.

Vapor Retarder
Mistakenly referred to as a vapor barrier. A material that resists the flow of water vapor.

Water-vapor Transmission
Water-vapor flow normal to two parallel surfaces of a material, through a unit area, under the conditions of a specified test such as ASTM E96.

American Society of Heating, Refrigerating and Air-Conditioning Engineers. 1997. *ASHRAE Handbook: Fundamentals.* Atlanta: ASHRAE (updated periodically).

_____. 1989. "Energy Efficient Design of New Buildings (Except New Low Rise Residential Buildings." *ASHRAE Standard 90.1-1989.* Atlanta: ASHRAE.

Asphalt Roofing Manufacturers Association. 1987. *Built-Up Roofing Systems Design Guide.* Calverton, Md.: ARMA.

Baker, Maxwell. 1980. *Roofs: Design, Application and Maintenance.* Sponsored by the National Research Council of Canada. Montreal: Multiscience Publications Limited.

Factory Mutual Research Corporation. *Factory Mutual System: A Guide to Equipment, Materials & Services Approved by Factory Mutual Research for Property Conservation.* Norwood, Mass: FMRC (updated annually and supplemented periodically through the year).

_____. "Loss Prevention Data Sheets." *Technical Bulletins.* Norwood, Mass: FMRC (dates vary).

Geary, Donald. 1978. *Roofs and Siding: A Practical Guide.* Reston, Va.: Reston Publishing.

Griffin, C.W., and R.L. Fricklas. 1996. *The Manual of Low-Slope Roof Systems.* 3d ed. New York: McGraw-Hill.

NAHB Research Foundation. 1979. *Insulation Manual.* 2nd ed. Rockville, Md.: NAHB Research Foundation, now known as NAHB Research Center (out of print).

National Roofing Contractors Association. 1996. *The NRCA Roofing and Waterproofing Manual.* 4th ed. 2 vols plus technical bulletins. Rosemont, Ill.: NRCA.

_____. 1988. *Application Techniques for Glass Fiber Roofing Felts.* By Richard Baxter. Rosemont, Ill.: NRCA.

Ramsey, Charles George. 1994. *Ramsey/Sleeper Architectural Graphic Standards.* 9th edition. John Ray Hoke, Jr. FAIA, ed. Somerset, NJ: John Wiley & Sons.

_____. 1998. *Ramsey/Sleeper Architectural Graphic Standards.* Cumulative supplement. John Ray Hoke, Jr. FAIA, ed. Somerset, NJ: John Wiley & Sons.

Revere Copper Products, Inc. 1982. *Copper & Common Sense: Sheet copper design principles and construction techniques.* 7th edition. Rome, N.Y.: Revere.

Sheet Metal and Air Conditioning Contractors' National Association, Inc. 1993. *Architectural Sheet Metal Manual.* 5th edition. Chantilly, Va: SMACNA.

Underwriters Laboratories Inc. 1998. *Roofing Materials and Systems Directory.* Northbrook, Ill.: UL (updated annually).

_____. 1998. *Fire Resistance Directory.* 2 vols. Northbrook, Ill.: UL (updated annually).

Watson, John. 1979. *Roofing Systems: Materials and Application.* Reston, Va.: Reston Publishing.

APA—The Engineered Wood Association (APA)
(formerly The American Plywood Association)
P.O. Box 11700
Tacoma, WA 98411
(253) 565-6600
web site: www.apawood.org

American Society for Testing and Materials (ASTM)
100 Barr Harbor Drive
West Conshohocken, PA 19428
(610) 832-9500
web site: www.astm.org

American Society of Civil Engineers (ASCE)
1801 Alexander Bell Drive
Reston, Virginia 20191
(800) 548-2723
web site: www.asce.org

American Society of Heating, Refrigerating and Air-Conditioning Engineers, Inc. (ASHRAE)
1791 Tullie Circle, N.E.
Atlanta, GA 30329
(404) 636-8400
web site: www.ashrae.org

Asphalt Roofing Manufacturers Association (ARMA)
4041 Powder Mill Road
Suite 404
Calverton, MD 20705
(301) 348-2002
web site: www.asphaltroofing.org

Factory Mutual Research Corporation (FM or FMRC)
1151 Boston-Providence Turnpike
Norwood, MA 02062
(781) 762-4300
web site: www.factorymutual.com

National Association of Home Builders Research Center
400 Prince George's Boulevard
Upper Marlboro, MD 20774
(301) 249-4000
web site: www.nahbrc.org

National Roofing Contractors Association (NRCA)
10255 West Higgins Road
Suite 600
Rosemont, IL 60018
(847) 299-9070
web site: www.nrca.net

Revere Copper Products, Inc.
P.O. Box 300
Rome, NY 13440
(315) 338-2022
web site: www.reverecopper.com

Roofing Industry Educational Institute (RIEI)
14 Inverness Drive East
Building H, Suite 110
Englewood, CO 80112
(303) 790-7200
web site: www.riei.org

Sheet Metal and Air Conditioning Contractors' National Association, Inc. (SMAC-NA)
4201 Lafayette Center Drive
Chantilly, VA 20151
(703) 803-2980
web site: www.smacna.org

Steel Deck Institute (SDI)
P.O. Box 25
Fox River Grove, IL 60021
(847) 462-1930
web site: www.sdi.org

Underwriters Laboratories Inc. (UL)
333 Pfingsten Road
Northbrook, IL 60062
(847) 272-8800
web site: www.ul.com

BUR final: 8/23/99 PAGES 13, 25, 44, 47

N C A R B

PROFESSIONAL DEVELOPMENT PROGRAM

LOW-SLOPE ROOFING I QUIZ

QUESTION 1

Insulating concrete should be used over which of the following types of deck to provide "downward" drying?

A. wire mesh and form board
B. pressure-treated plywood
C. venting galvanized steel
D. cast-in-place structural concrete

QUESTION 2

An ideal insulation should have which of the following properties?

A. tough, wetting resistant, low cost, with low U-value
B. lightweight, fire resistant, low cost, with low *k value*
C. dimensionally stable, lightweight, tough, with wetting resistance
D. high-temperature resistant, fire resistant, impact resistant, with low U-value

QUESTION 3

Insulation delivered to a construction site must be stored above ground and

A. covered with tarpaulins.
B. in plastic shipping wrappers.
C. on end to prevent flat spots.
D. in a designated storage area.

QUESTION 4

Which of the following is the most important condition for the longevity of a roof?

A. secure seams
B. good drainage
C. few penetrations
D. adequate dividers

QUESTION 5

To control thermal expansion in extruded metal (heavy metal) gravel stops, an architect should specify a maximum length of approximately which of the following?

A. 6 feet
B. 8 feet
C. 10 feet
D. 14 feet

QUESTION 6

The technique called *double pouring* refers to

A. surfacing applied twice.
B. combining asphalt and bitumen.
C. two coats of asphalt under last felt.
D. a double coating between layers of insulation.

QUESTION 7

The equiviscous temperature (EVT) refers to which of the following?

A. the temperature to achieve ideal bonding
B. the temperature when flash point is reached
C. the temperature when roofing asphalt breaks down
D. the minimum temperature for cold weather application

QUESTION 8

When applied to a roof, the term *blow off* describes when

A. the roof is lifted up.
B. the roofing is alligatored.
C. ponded water is blown off.
D. a ballast becomes airborne.

QUESTION 9

An advantage of cast-in-place concrete over precast concrete is that it

A. cures more rapidly.
B. can be sloped to a drain.
C. is much lighter in weight.
D. does not require stripping.

QUESTION 10

Drains on low-slope roofs should be installed in sumps of at least

A. 24 inches square.
B. 36 inches square.
C. 40 inches square.
D. 48 inches square.

QUESTION 11

Which of the following is the most likely cause of blistering roof felts?

A. gravel depth
B. number of plies
C. ultraviolet rays
D. moisture entrapment

QUESTION 12

Underwriters Laboratories (UL) fire tests are performed on

A. membrane density.
B. decking stiffness.
C. separate components.
D. complete assemblies.

QUESTION 13

Which of the following will minimize separation of the membrane from the roof deck?

A. minimal deflection
B. elimination of ponding
C. increased number of felts
D. reduced penetration of moisture

QUESTION 14

Which of the following is the most likely cause of condensation within insulation?

A. The outside temperature is lower than the inside temperature.
B. The outside temperature is higher than the inside temperature.
C. The dew-point temperature is lower than the vapor retarder temperature.
D. The dew-point temperature is higher than the vapor retarder temperature.

QUESTION 15

The primary function of a roof drain with a flow weir is to

A. connect the roof drain to the scupper.
B. allow the scupper to be less than 2 inches.
C. control the release of accumulated roof water.
D. allow the overflow to drain into a separate plumbing system.

QUESTION 16

A fastener for a extruded-metal gravel stop is most likely to shear due to which of the following?

A. use of 18-gauge metal
B. the length is more than 15 feet
C. cover plates are secured with mastic
D. the way the metal is attached to the nailer

QUESTION 17

Water penetration rarely occurs

A. in the plane of the roof.
B. in rooftop expansion joints.
C. at terminations of gravel stops.
D. at terminations of rising walls and parapets.

QUESTION 18

Three-pound sheet lead is used to flash vent stacks because it can be

A. turned down into a pipe.
B. obtained at a low cost.
C. resistant to corrosion.
D. sealed into the roofing membrane.

QUESTION 19

A hood should be used for counterflashing over metal base flashing when

A. snow drifts can exceed 24 inches.
B. a pipe penetration is 4 inches in diameter.
C. wind forces frequently exceed 25 miles per hour.
D. the surface temperature of the penetration is 120 F.

QUESTION 20

Roofing felts frequently rupture due to

A. decks having very low slopes.
B. occupancies having high humidity.
C. differential movement of components.
D. the use of various bituminous products.

QUESTION 21

A surface aggregate should have which of the following characteristics?

1. translucent and light in color
2. fall within a size range of No. 6 or No. 67
3. rounded smooth surfaces
4. sandy consistency for embedment

A. 1 and 2 only
B. 1 and 4 only
C. 2 and 3 only
D. 3 and 4 only

QUESTION 22

The minimum spacing for roof insulation fasteners is

A. one per square foot.
B. one per two square feet.
C. 12 inches on center each way.
D. 16 inches on center each way.

QUESTION 23

Which of the following conditions are critical for proper adhesion between felt plies?

1. temperature of the bitumen
2. outdoor temperature
3. uniform film of bitumen
4. contact between felts

A. 1 and 3 only
B. 1 and 4 only
C. 2 and 3 only
D. 2 and 4 only

QUESTION 24

A concrete framed roof structure is to be designed to provide the minimum slope required by the model building codes. What is the height difference between the top beams on combined spans of 32 feet each with a constant slope from side to side?

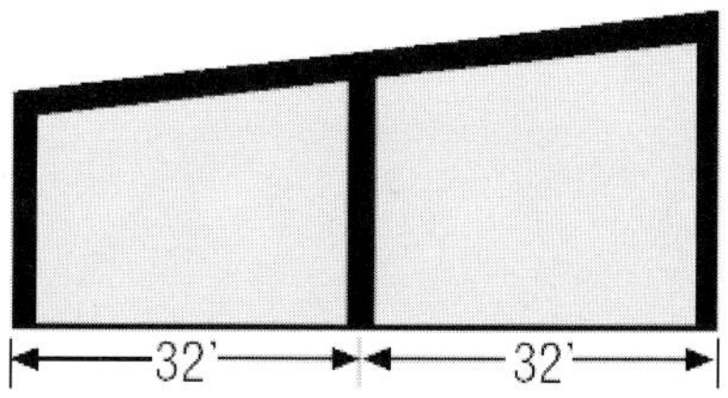

A. 8 inches
B. 12 inches
C. 16 inches
D. 24 inches

QUESTION 25

What type of asphalt should be used on the roof shown here?

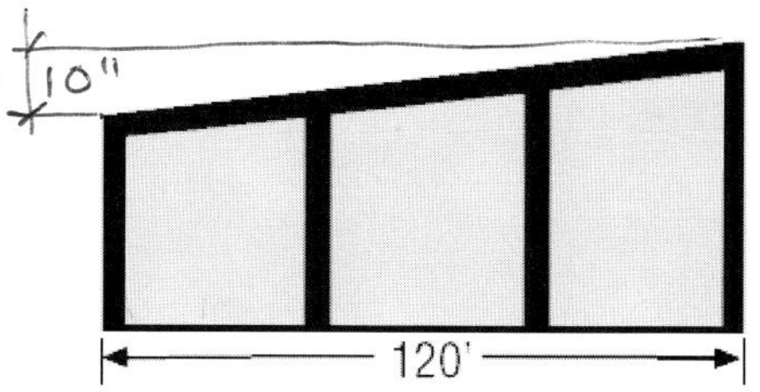

A. I and II only
B. I and III only
C. II and IV only
D. III and IV only

QUESTION 26

This detail represents which of the following?

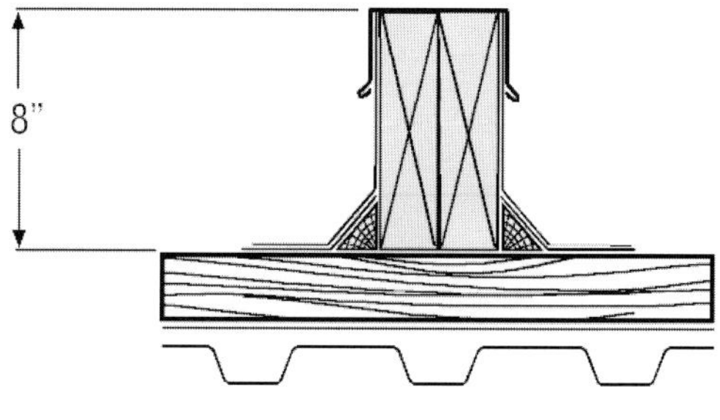

A. area divider
B. pitch pocket
C. equipment curb
D. expansion joint

QUESTION 27

Which of the following should be used to prevent condensation within the BUR?

A. impervious cap sheet
B. two layers of insulation staggered
C. vapor retarder on bottom of insulation
D. roof system vents within the insulation

QUESTION 28

The greatest number of fasteners holding the insulation to the roof should occur at the

A. cants.
B. curbs.
C. center.
D. corners.

QUESTION 29

What gauge metal deck is most commonly used where the deck serves as a diaphragm?

A. 18 gauge
B. 20 gauge
C. 22 gauge
D. 24 gauge

QUESTION 30

How many roof drains are required on a flat roof which is 160 feet by 240 feet, if no drain is farther than 40 feet from the edge of the roof?

A. 4
B. 6
C. 8
D. 10

QUESTION 31

Overflow drains are required to prevent

A. roof storm-water overloads.
B. water damage to parapet walls.
C. water freezing around deck drains.
D. overloading of the building plumbing system.

QUESTION 32

During inspection, it is observed that the mechanical fasteners attached to the roof deck are not secure. Which of the following is the likely cause?

A. The specifications called for metal deck.
B. The deck was made of cast-in-place concrete.
C. The wrong type of mechanical fasteners were used.
D. The cementitious wood-fiber thickness was 4 inches.

QUESTION 33

The asphalt's flash point is 500 F and the equiviscous temperature (EVT) is 375 F. What is the minimum temperature when the asphalt is applied?

A. 325 F
B. 350 F
C. 375 F
D. 425 F

QUESTION 34

Which of the following occurs as the R-value of insulation increases?

A. The roof decking increases in temperature.
B. There is a reduction in thermal efficiency.
C. Roof-surface temperature variations are increased.
D. There is less possibility of movement in the system.

QUESTION 35

The two-layer system on metal decks refers to

A. the application of adhesive.
B. the attachment of insulation.
C. a procedure for installing felts.
D. applying stripping over felt joints.

QUESTION 36

Moisture entering a roof system by capillary action may be due to

A. melting snow drifts.
B. the rigidity of base flashing.
C. saturation of rigid roof insulation.
D. penetration of moisture through coping joints.

QUESTION 37

The overall coefficient of thermal transmittance (U) for a complete system, such as a roof, is calculated by using which of the following formulas?

A. $U = \left(\frac{1}{C_1 + C_2 + C_3 \ldots}\right)$

B. $U = \left(\frac{1}{k_1 + k_2 + k_3 \ldots}\right)$

C. $U = \left(\frac{1}{R_1 + R_2 + R_3 \ldots}\right)$

D. $U = \left(\frac{1}{Btu_1 + Btu_2 + Btu_3 \ldots}\right)$

QUESTION 38

Roof drain sumps are best formed by which of the following?

A. cast-in-sleeve
B. roofing cement
C. tapered insulation
D. hole cut into the deck

QUESTION 39

When using a non-combustible grouted system prior to roofing, the effects of moisture released from the drying grout can best be minimized by

A. mopping the first base sheet.
B. spray applying a drying agent.
C. mechanically fastening the base sheet.
D. using an adhesive to fasten a fiberglass sheet.

QUESTION 40

Cementitious wood-fiber decks offer all of the following properties EXCEPT

A. finished ceilings.
B. thermal insulation.
C. acoustical insulation.
D. resistance to humidity.

QUESTION 41

What is the minimum dimension from the toe of a cant to a roof penetration?

A. 6 inches
B. 12 inches
C. 18 inches
D. 24 inches

QUESTION 42

The most important function of light-colored surfacings is to

A. reduce solar reflection.
B. reduce surface temperature.
C. resist damage from foot traffic.
D. improve water tightness of the system.

QUESTION 43

Which of the following are basic components of a BUR membrane?

1. reflective material
2. adhesive
3. reinforcement
4. surfacing

A. 1, 2, and 3 only
B. 1, 3, and 4 only
C. 2, 3, and 4 only
D. 1, 2, 3, and 4

QUESTION 44

Lightweight insulating deck and cementitious panels should NOT be used

A. in cold climates.
B. on low-pitched roofs.
C. where subject to high winds.
D. over areas of high humidity.

QUESTION 45

The main reason that prestressed concrete decks do NOT provide a satisfactory surface to receive roofing components is because

A. the camber is not uniform.
B. the surfaces are not smooth.
C. long spans are subject to flexing.
D. they have a high moisture absorption rate.

QUESTION 46

When the temperature of a roof surface varies within the range of -20 F to 160 F, the roof components

A. move uniformly.
B. expand similarly.
C. contract equally.
D. move differentially.

QUESTION 47

Moisture in the insulation

A. increases thermal efficiency.
B. decreases thermal efficiency.
C. increases the rate of expansion.
D. decreases the rate of expansion.

QUESTION 48

If the roof surface temperature varies as much as 180 degrees, the architect should design the roof system to control

A. movement.
B. fishmouths.
C. blistering.
D. condensation.

QUESTION 49

Which of the following should be done first to prevent moisture migration from the building interior to the insulation?

A. Locate the dew point in the roof membrane.
B. Eliminate installation of the vapor retarder.
C. Increase the R-value of the insulation below the deck.
D. Ensure that the vapor retarder is warmer than the dew-point temperature.

QUESTION 50

Which of the following is the most likely cause of water ponding on a roof?

A. location of the drains
B. deflections of the decking
C. overflows are set 6 inches high
D. roof slopes of 1/8 inch per foot

QUESTION 51

If the cant strip at the parapet was omitted during construction, the base flashing will most likely

A. crack.
B. shear.
C. blister.
D. fishmouth.

QUESTION 52

The primary reason that roof drains are temporarily plugged during construction is to

A. test the overflow drains.
B. avoid clogging from debris.
C. evaluate the roofing assembly.
D. prevent water from entering the building.

QUESTION 53

The total minimum dimension of base flashing, when properly installed horizontally over the membrane and vertically above the cant, would be approximately

A. 4 inches.
B. 8 inches.
C. 12 inches.
D. 16 inches.

QUESTION 54

Penetrations through roofing membranes are best flashed with

A. pitch pans.
B. pitch pockets.
C. base flashing and counterflashing.
D. concealed lead-coated copper.

QUESTION 55

The primary function of felts is to

A. add insulating value to the roof system.
B. provide reinforcing in the roof membrane.
C. cover and bridge over rough deck surfaces.
D. reduce the amount of bituminous material required.

QUESTION 56

What is the approximate proper exposure for three plies of 36-inch-wide felt in a BUR?

A. 11 3/8 inches
B. 12 1/4 inches
C. 15 3/4 inches
D. 18 1/2 inches

QUESTION 57

Which of the following rooftop equipment design options should be considered for a building that houses manufacturing processes?

A. Use cap sheets on the roof surface.
B. Layer metal reinforcement into the roof.
C. Coat the roof with a light-color surface.
D. Coat the roof with an extra thick surface.

QUESTION 58

Which of the following is standard practice in roofing design?

1. Keep door sills 6 inches above the roof plane.
2. Separate curbed openings by 10 inches horizontally.
3. Extend base flashing 8 inches to 12 inches up vertical walls.
4. Use pressure-treated wood as nailers and cants.

A. 1 and 2 only
B. 1 and 3 only
C. 2 and 4 only
D. 3 and 4 only

QUESTION 59

Which of the following locations requires base flashing?

1. masonry parapets 16 inches above roof
2. along expansion joints
3. around roof penetrations
4. in built-in gutters

A. 1 and 4 only
B. 2 and 3 only
C. 1, 2, and 3 only
D. 2, 3, and 4 only

QUESTION 60

Water ponding on a roof may enter the roof system because of

A. permeability.
B. surface tension.
C. negative pressure.
D. hydrostatic pressure.

QUESTION 61

Which of the following has an effect on membrane slippage?

1. type of membrane
2. equiviscous temperature (EVT)
3. quantity of asphalt
4. air temperature

A. 1 and 2 only
B. 1 and 4 only
C. 2 and 3 only
D. 3 and 4 only

QUESTION 62

The top of base flashings should be secured to a masonry wall by which of the following?

A. hot-applied bitumen
B. cold-applied adhesive
C. torched modified bitumen
D. nailing at 8 inches on center

R-value

QUESTION 63

Outside Air	**.17**
BUR	**.33**
Metal Deck	**.00**
Air Space	**.93**
Ceiling	**1.56**
Interior Air Film	**.61**

If a required R-value for a roof system is 22, what is the R-value for the insulation in the above system?

A. 12.4
B. 16.0
C. 18.4
D. 20.0

QUESTION 64

Side laps on steel decking shall be secured at the maximum spacing of

A. 18 inches.
B. 24 inches.
C. 30 inches.
D. 36 inches.

QUESTION 65

Gypsum concrete decks are a mixture of water,

A. wood chips and gypsum.
B. fine aggregate and gypsum.
C. portland cement and gypsum.
D. wood chips and portland cement.

QUESTION 66

A roofing project cannot be completed in one day. What must be done to protect the unfinished roof?

A. Place cap sheets on the felts.
B. Place plywood sheets over the felts.
C. Temporarily place aggregate on the felts.
D. Glaze the felts with a thin coat of asphalt.

QUESTION 67

During inspection of the precast concrete roof deck, the camber varies as much as 1 inch. After reviewing structural data, the architect might recommend

A. replacing the deck.
B. sanding the surface.
C. placing grout between the forms.
D. adding 2 inches of concrete topping.

QUESTION 68

Polystyrene foam insulation can be attached by

A. cold-tar pitch.
B. mechanical fasteners.
C. 225 F degrees hot bitumen.
D. solvent-based adhesives.

QUESTION 69

Roof insulation has which of the following drawbacks?

1. Stores heat energy, accelerating membrane deterioration.
2. Increases summer temperatures on membrane.
3. Can store moisture, increasing deterioration.

A. 1 and 2 only
B. 1 and 3 only
C. 2 and 3 only
D. 1, 2, and 3

QUESTION 70

A proper retarder is placed between the

A. felt and cap sheet.
B. insulation and felt.
C. deck and rigid insulation.
D. deck and structural frame.

QUESTION 71

Which of the following types of insulation should be used where high roof traffic will occur?

A. Perlite-fiber
B. glass-fiber board
C. polystyrene foam
D. wood-fiber board

QUESTION 72

Surface-mounted cap flashing on a tilt-up concrete wall should be sealed at the top with

A. butyl caulking.
B. thiokol caulking.
C. silicone caulking.
D. urethane caulking.

QUESTION 73

Concealed flashings formed from self-adhering modified bitumens, laminated metal foils, EPDM, and PVC should not be used as cap flashings because

1. of poor UV resistance.
2. they will stain the finish wall.
3. excessive heat gain will occur.
4. of their inability to resist wind uplift.

A. 1 and 4 only
B. 2 and 3 only
C. 2 and 4 only
D. 3 and 4 only

QUESTION 74

When plastic drains are used, one condition that the roofer must be aware of is that the

A. drain is not grounded.
B. ballast may clog the drain.
C. drain may soften when applying MBR.
D. building codes may not allow their use.

QUESTION 75

Flashing masonry to bituminous products requires the surface to be

A. primed.
B. scarfed.
C. sealed.
D. filled.